Table of Contents

List of Exhibits

Acronyms and Abbreviations

APFBC	Advanced pressurized fluidized bed combustion
ASTM	American Society for Testing and Materials
ASME	American Society of Mechanical Engineers
ASU	Air separation unit
BACT	Best Available Control Technology
Btu	British thermal unit
CAA	Clean Air Act
CCS	Carbon capture and storage
CEMS	Continuous emission monitoring system
CFB	Circulating fluidized bed
CH_4	Methane
CO	Carbon monoxide
CO_2	Carbon dioxide
EGU	Electric generating unit
EPA	U.S. Environmental Protection Agency
FBC	Fluidized bed combustion
EPRI	Electric Power Research Institute
FGD	Flue gas desulfurization
GHG	Greenhouse gas
H_2O	Water
HRSG	Heat recovery steam generator
HHV	Higher heating value
IGCC	Integrated gasification combined cycle
IEA	International Energy Agency
kJ	Kilojoule
kW	Kilowatt
kWh	Kilowatt-hour
LCOE	Levelized cost of electricity
Mg	Megagram
MMBtu/hr	Million Btu per hour
MPa	Megapascal

MW	Megawatt
MWe	Megawatt electrical
MWh	Megawatt-hour
MSW	Municipal solid waste
N_2O	Nitrous oxide
NETL	National Energy Technology Laboratory
NO_X	Nitrogen oxides
O&M	Operation and maintenance
PC	Pulverized coal
PFBC	Pressurized fluidized bed combustion
PM	Particulate matter
PRB	Power River Basin
scfm	Standard cubic feet per minute
SO_2	Sulfur dioxide
SO_3	Sulfur trioxide
SNCR	Selective noncatalytic reduction
ton/day	tons per day
ton/yr	tons per year
U.S. DOE	U.S. Department of Energy
U.S. EIA	U.S. Energy Information Administration

1. Introduction

This document is one of several white papers that summarize readily available information on control techniques and measures to mitigate greenhouse gas (GHG) emissions from specific industrial sectors. These white papers are solely intended to provide basic information on GHG control technologies and reduction measures in order to assist States and local air pollution control agencies, tribal authorities, and regulated entities in implementing technologies or measures to reduce GHGs under the Clean Air Act, particularly in permitting under the prevention of significant deterioration (PSD) program and the assessment of best available control technology (BACT). These white papers do not set policy, standards or otherwise establish any binding requirements; such requirements are contained in the applicable EPA regulations and approved state implementation plans.

This document provides information on control techniques and measures that are available to mitigate GHG emissions from the coal-fired electric generating sector at this time. The primary GHG emitted by the coal-fired electric generation industry is carbon dioxide (CO_2), and the control technologies and measures presented in this document focus on this pollutant. While a large number of available technologies are discussed here, this paper does not necessarily represent all potentially available technologies or measures that that may be considered for any given source for the purposes of reducing its GHG emissions. For example, controls that are applied to other industrial source categories with exhaust streams similar to the cement manufacturing sector may be available through "technology transfer" or new technologies may be developed for use in this sector.

The information presented in this document does not represent U.S. EPA endorsement of any particular control strategy. As such, it should not be construed as EPA approval of a particular control technology or measure, or of the emissions reductions that could be achieved by a particular unit or source under review.

1.1   Electric Power Generation Using Coal

Electricity is generated at most electric power plants by using mechanical energy to rotate the shaft of electromechanical generators. The mechanical energy needed to rotate the generator shaft can be produced from the conversion of chemical energy by burning fuels or from nuclear fission; from the conversion of kinetic energy from flowing water, wind, or tides; or from the conversion of thermal energy from geothermal wells or concentrated solar energy. Electricity also can be produced directly from sunlight using photovoltaic cells or by using a fuel cell to electrochemically convert chemical energy into an electric current.

In 2008, approximately 70% of the electricity used in the United States was generated by burning fossil fuels (coal, natural gas, petroleum liquids) (U.S. EIA 2010). The combustion of a fossil fuel to generate electricity can be either: 1) in a steam generating unit (also referred to simply as a "boiler") to feed a steam turbine that, in turn, spins an electric generator: or 2) in a combustion turbine or a reciprocating internal combustion engine that directly drives the generator. Some modern power plants use a "combined cycle" electric power generation process, in which a gaseous or liquid fuel is burned in a combustion turbine that both drives electrical generators and provides heat to produce steam in a heat recovery steam generator

(HRSG). The steam produced by the HRSG is then fed to a steam turbine that drives a second electric generator. The combination of using the energy released by burning a fuel to drive both a combustion turbine generator set and a stream turbine generator significantly increases the overall efficiency of the electric power generation process.

Coal is the most abundant fossil fuel in the United States and is predominately used for electric power generation. In 2008, approximately 49% of the net electricity generated in the U.S. was produced by coal (U.S. EIA 2010). Historically, electric utilities have burned solid coal in steam generating units. However, coal can also be first gasified and then burned as a gaseous fuel. The integration of coal gasification technologies with the combined cycle electric generation process is called an integrated gasification combined cycle (IGCC) system or a "coal gasification facility". For the remainder of this document, the term "electric generating unit" or "EGU" is used to mean a solid fuel-fired steam generating unit that serves a generator that produces electricity for sale to the electric grid.

2. Coal-Fired Electric Generating Units

This section provides a summary overview of the types or ranks of coal that are typically burned in EGUs operating in the United States, the most commonly used combustion processes, and the resulting emissions of greenhouse gases.

2.1 Coals Burned in U.S. EGUs

In the United States, coals are ranked based on the degree of metamorphism (effectively, the geological age of the coal and the conditions under which the coal formed). These classification criteria have been standardized by the American Society for Testing and Materials (ASTM) method D-388. Under the ASTM method, coals are divided into four major categories called "ranks:" anthracite, bituminous coal, subbituminous coal, and lignite. Typical coal characteristics for the three most commonly used coal ranks are summarized in **Exhibit 2-1**.

Exhibit 2-1. Selected characteristics of major coal ranks used for electricity generation in the United States.

Coal Rank[a]	Higher Heating Value (HHV) Range Defined by ASTM D-388	Typical Coal Moisture Content[b]	Coal Delivered for U.S. Electric Power Production in 2008[c,d]		
			Total Coal Quantity Delivered Nationwide (1,000 tons)	Average Ash Content	Average Sulfur Content
Bituminous	>10,500 Btu/lb	2 to 16%	463,943	10.6%	1.68%
Subbituminous	<10,500 Btu/lb and >8,300 Btu/lb	15 to 30%	522,228	5.8%	0.34%
Lignite	< 8,300 Btu/lb	25 to 40%	68,945	13.8%	0.86%

[a] Anthracite coal use is limited to reclaiming coal from coal refuse piles for use in a few power plants located close to the anthracite mines in eastern Pennsylvania.
[b] Reference: U.S. EPA, 2001.
[c] Reference: U.S. EIA, 2010, Table 3.6.
[d] Includes data collected from electric utilities, independent power producers, and combined heat and power producers.

Most coal-fired EGUs in the United States burn either bituminous or subbituminous coals. Approximately one half of the tonnage of coals delivered to U.S. electric power generation facilities was subbituminous (49.5%), and another 44% was bituminous coal. Some coal-fired EGUs burn multiple coal ranks. At many of these facilities, the coals are blended together before firing. However, some facilities may switch between coal ranks because of site-specific considerations. The largest sources of bituminous coals burned in EGUs are mines in regions along the Appalachian Mountains, in southern Illinois, and in Indiana. Additional bituminous coals are supplied from mines in Utah and Colorado. The vast majority of subbituminous coals are supplied from mines in Wyoming and Montana, and many EGUs burn subbituminous coals from the Powder River Basin (PRB) region in Wyoming. This material is often referred to simply as "PRB coal."

In general, the burning of lignite or anthracite by electric utilities is limited to those EGUs that are located near the mines supplying the coal. Lignite accounted for approximately 6.5% of the total tonnage of coal delivered to electric utility power plants in 2008. All of those facilities were located near the coal deposits from which the lignite was mined in Texas, Louisiana, Mississippi, Montana, or North Dakota. Similarly, anthracite use was limited to a few power plants located close to the anthracite mines in eastern Pennsylvania. The coal-fired EGUs at those facilities primarily burn anthracite that has been reclaimed from coal refuse piles of previous mining operations. In general, "coal refuse" means any by-product of coal mining or coal cleaning operations with an ash content greater than 50 % (by weight) and a heating value less than 13,900 kilojoules per kilogram (kJ/kg) (6,000 Btu per pound (Btu/lb) on a dry basis. Coal refuse piles from previous mining operations are primarily located in Pennsylvania and West Virginia. Current mining operations generate less coal refuse than older ones.

2.2   Coal Utilization in U.S. EGUs

Steam turbine power plants operate on the Rankine thermodynamic cycle. The steam is produced by the boiler, where water pumped into the boiler ("feedwater") passes through a series of tubes to capture heat released by coal combustion and then boils under high pressure to become superheated steam. The superheated steam leaving the boiler then enters the steam turbine throttle, where it powers the turbine and connected generator to make electricity.

After the steam expands through the turbine, it exits the back end of the turbine into the surface condenser, where it is cooled and condensed back to water. This condensate is then returned to the boiler through high-pressure feed pumps for reuse. Heat from the condensing steam is normally rejected to cooling water circulated through the condenser which then goes to a surface water body, such as a river, or to an on-site cooling tower.

An EGU can be classified as either dry or wet bottom, depending on the ash removal technique used. Dry bottom boilers fire coals with high ash fusion temperatures, allowing for solid ash removal. In the less common wet bottom (slag tap) boilers, coal with a low ash fusion temperature is fired, and molten ash is drained from the bottom of the boiler.

To improve the overall thermal conversion efficiency of the Rankine cycle, the majority of EGUs include a series of heat recovery sections. These sections are located downstream from the furnace chamber and are used to extract additional heat from the flue gas. The first section contains a "superheater," which is used to increase the steam temperature. The second heat recovery section contains a "reheater," which reheats the steam exhausted from the first stage of the steam turbine. This steam is then returned for another pass thorough a second stage of the turbine. The reheater is followed by an "economizer," which preheats the condensed feedwater recycled back to the boiler tubes in the furnace. The final heat recovery section is the "air heater," which preheats the ambient air used for coal combustion. The flue gas exhausted from the boiler passes through particulate matter (PM) and other air emissions control equipment before being vented to the atmosphere through a stack.

Coal-fired EGUs use one of five basic coal utilization processes.

- Stoker-fired
- Pulverized coal (PC)
- Cyclone-fired
- Fluidized-bed combustion (FBC)
- Coal gasification (IGCC)

Pulverized coal is the coal-firing configuration predominately used at existing U.S. electric utility power plants, and is also most frequently selected for new coal-fired EGU projects. Fluidized-bed combustion and coal gasification are newer technologies that, depending on project specific requirements, can be considered as alternatives to building a new PC-fired EGU. Cyclone and stoker firing are older technologies that are generally not considered when building new coal-fired EGUs. However, some existing cyclone and stoker-fired units are still in operation. The characteristics of each of the coal-firing configurations are summarized in **Exhibit 2-2** and discussed further in the following sections.

2.2.1 Stoker-Fired Coal Combustion

First introduced to the electric utility industry in the late 1800s, stoker-fired coal combustion is the oldest boiler coal-firing design. In a stoker-fired boiler, the coal is crushed and burned on a grate. Heated air passes upward through openings in the grate. Stokers are classified according to the way coal is fed to the grate – as underfeed stokers, overfeed stokers, and spreader stokers (see **Exhibit 2-2**). Stoker firing coal combustion is an obsolete technology for new coal-fired EGUs because the other newer coal combustion technologies provide superior coal combustion efficiency, applicability, and other advantages. There are still a few small stoker-fired EGUs in service in the U.S., but as these units are retired no new coal-fired stoker-fired EGUs are expected to be built. The majority of new stoker-fired boiler capacity is expected to occur at municipal solid waste combustor facilities and facilities burning solid biomass.

2.2.2 Pulverized-Coal Combustion

Pulverizing coal into a very fine powder allows the coal to be burned more easily and efficiently. For a PC-fired EGU, the coal must first be pulverized in a mill to the consistency of talcum powder (i.e., at least 70% of the particles will pass through a 200-mesh sieve). The pulverized coal is generally entrained in primary combustion air before being blown through the burners into the combustion chamber where it is fired in suspension. PC-fired boilers are classified by the firing position of the burners either as wall-fired or tangential-fired (see **Exhibit 2-2**).

A PC-fired boiler consists of multiple sections, and **Exhibit 2-3** presents a simplified schematic of the major components of a PC-fired boiler using subcritical steam conditions. The pulverized coal is ignited and burned in the section of the boiler called the "furnace chamber" (or sometimes the "firebox"). Ambient air blown into the furnace chamber provides the oxygen required for combustion. The walls of the furnace chamber are lined with vertical tubes containing the feedwater. Heat transfer from the hot combustion gases in the furnace boils the water in the tubes to produce the high-temperature, high-pressure steam. The steam is separated

Exhibit 2-2. Characteristics of coal-firing configurations used for U.S. EGUs.

Coal-firing Configuration	Application to U.S. EGUs	Coal Combustion Process Description	Distinctive Design/Operating Characteristics	
Stoker-fired	▪ Oldest coal-firing design first introduced to the electric utility industry in the late 1800s. ▪ Not a significant contributor to overall U.S. nationwide MW generating capacity. ▪ New EGUs are not expected to use this coal-firing design because of the superior performance and advantages of newer coal combustion technologies.	Coal is crushed into large lumps and burned in a fuel bed on a moving, vibrating, or stationary grate. Coal is pushed, dropped, or thrown onto the grate by a mechanical device called a "stoker."	Spreader-stoker	A flipping mechanism throws the coal into the furnace above the grate. The fine coal particles burn in suspension while heavier coal lumps fall to the grate and burn in a fuel bed.
			Underfeed	Coal fed by pushing the coal up underneath the burning fuel bed.
			Traveling grate	Coal is fed by gravity onto a moving grate and leveled by a stationary bar at the furnace entrance.
Pulverized-Coal Combustion	▪ Coal-firing design predominately used at existing U.S. EGUs. ▪ In 2008, consumed ~ 92% of total coal consumed by U.S. EGUs.[a] ▪ Currently coal-firing design of choice for new large coal-fired EGUs (> 400 MWe) built in U.S.	Coal is ground to a fine powder that is pneumatically fed to a burner where it is mixed with combustion air and then blown into the furnace. The pulverized-coal particles burn in suspension in the furnace. Unburned and partially burned coal particles are carried off with the flue gas.	Wall-fired	An array of burners fire into the furnace horizontally, and can be positioned on one wall or opposing walls depending on the furnace design.
			Tangential-fired (Corner-fired)	Multiple burners are positioned in opposite corners of the furnace producing a fireball that moves in a cyclonic motion and expands to fill the furnace.
Cyclone	▪ Existing cyclone EGUs in U.S. constructed prior to 1981. ▪ In 2008, consumed ~ 6% of total coal consumed by U.S. EGUs. ▪ New EGUs are not expected to use this boiler type because of the commercial availability of FBC technology.	Coal is crushed into small pieces and fed through a burner into the cyclone furnace. A portion of the combustion air enters the burner tangentially creating a whirling motion to the incoming coal.	Designed to burn coals with low-ash fusion temperatures that are difficult to burn in PC boilers. The majority of the ash is retained in the form of a molten slag.	

Exhibit 2-2. Continued.

Coal-firing Configuration	Application to U.S. EGUs	Coal Combustion Process Description	Distinctive Design/Operating Characteristics	
Fluidized-bed Combustion	▪ FBC EGUs increasingly being built in U.S. to burn low rank coals, coal refuse, and blends of coal with other solid fuels such as petroleum coke or biomass. ▪ In 2008, consumed approximately 2% of total coal consumed by U.S. EGUs.[a] ▪ Atmospheric FBC EGUs are currently operating in the U.S. with generating capacities in the range of 250 to 300 MWe. ▪ No Pressurized FBC boilers currently used for U.S. EGUs	Coal is crushed into fine particles. The coal particles are suspended in a fluidized bed by upward-blowing jets of air. The result is a turbulent mixing of combustion air with the coal particles. Typically, the coal is mixed with a sorbent such as limestone (for SO_2 emission control). The unit can be designed for combustion within the bed to occur at atmospheric or elevated pressures. Operating temperatures for FBC are in the range of 1,500 to 1,650°F (800 to 900°C).	Bubbling fluidized bed (BFB)	Operates at relatively low gas stream velocities and with coarse-bed size particles. Air in excess of that required to fluidize the bed passes through the bed in the form of bubbles.
			Circulating fluidized bed (CFB)	Operates at higher gas stream velocities and with finer-bed size particles. No defined bed surface. Must use high-volume, hot cyclone separators to recirculate entrained solid particles in flue gas to maintain the bed and achieve high combustion efficiency.
Coal Gasification (e.g., IGCC)	▪ Limited application to EGUs to date. ▪ Some new proposed EGU projects using coal gasification as part of IGCC plant.	Synthetic combustible gas ("syngas") derived from an on-site coal gasification process is burned in a combustion turbine. The hot exhaust gases from the combustion turbine pass through a heat recovery steam generator to produce steam for driving a steam turbine/generator unit.	Coal gasification units are unique from the other coal-firing configurations because a gaseous fuel (synfuel or syngas) is burned instead of solid coal and combines the Rankine and Brayton thermodynamic cycles as is the case for a combined cycle power plant.	

[a] Source: U.S. EIA, 2008.

11

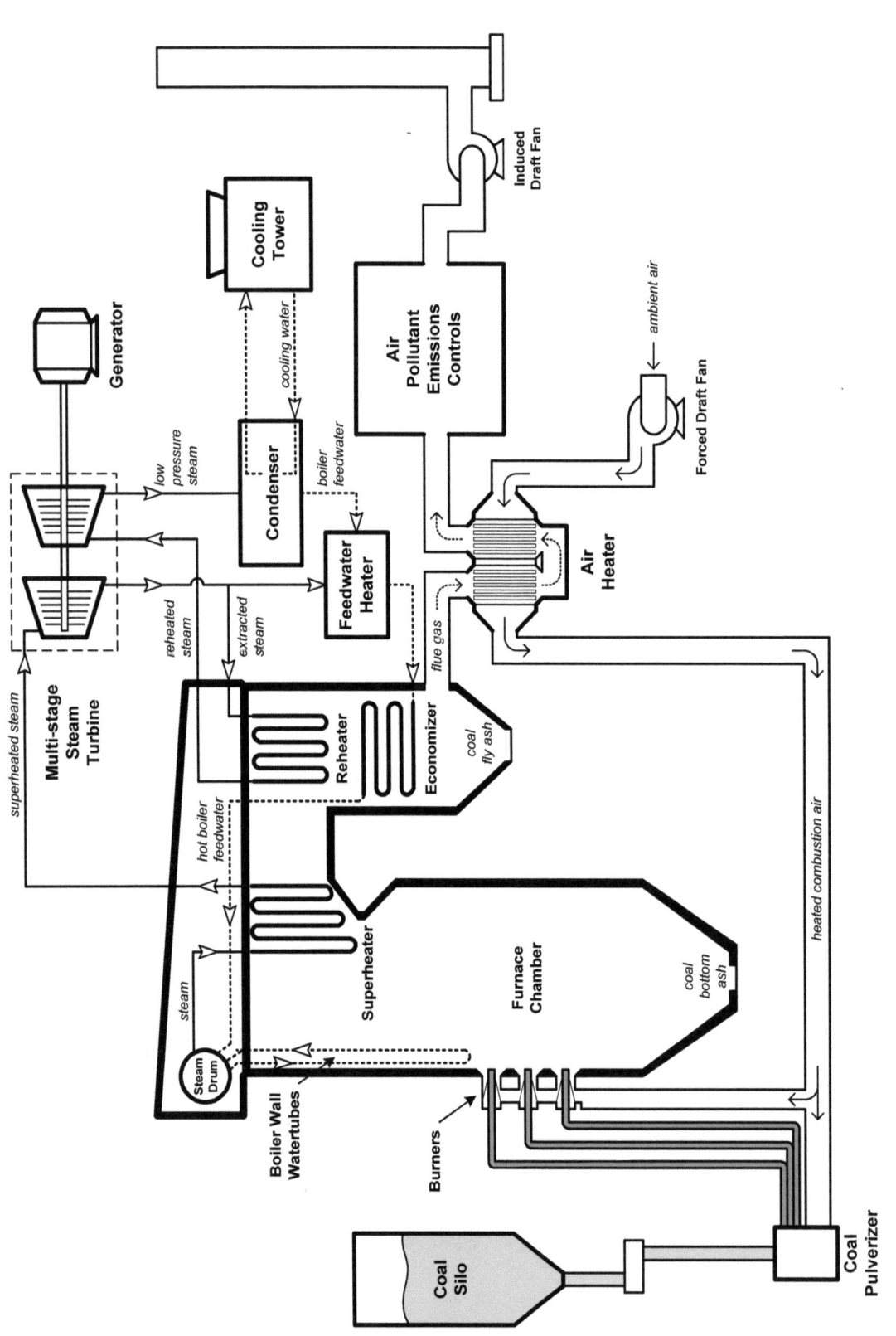

Exhibit 2-3. Simplified schematic of a PC-fired EGU using a subcritical boiler.

12

from boiler water in a steam drum and sent to the steam turbine. The remaining water in the drum re-enters the boiler for further conversion to steam. The hot combustion products are vented from the furnace in a gas stream called collectively flue gas.

2.2.3 Cyclone Coal Combustion

Cyclone coal combustion technology was developed as an alternative to PC-firing because it requires less pre-processing of the coal and allows for the burning of lower rank coals with higher moisture and ash contents. Cyclone boilers use burner design and placement (i.e., several water-cooled horizontal burners) to produce high-temperature flames that circulate in a cyclonic pattern. The coal is crushed to a 4-mesh size, and then fed tangentially with primary air, to a horizontal cylindrical combustion chamber. In this chamber, small coal particles are burned in suspension, while the larger particles are forced against the outer wall. The high temperatures developed in the relatively small boiler volume, combined with the low fusion temperature of the coal ash, causes the ash to form a molten slag, which is drained from the bottom of the boiler through a slag tap opening. Existing cyclone EGUs in the U.S. were designed or installed before 1981. Cyclone EGUs have high nitrogen oxides (NO_X) emission rates and no new cyclone boilers are expected to be built. Fluidized-bed combustion is an alternative technology that is able to burn lower rank coals without high NO_X emissions.

2.2.4 Fluidized-Bed Combustion

The term "fluidized" refers to the state of the bed materials (fuel and inert material [or sorbent]) as gas passes through the bed. In a typical FBC EGU, combustion occurs when coal and a sorbent, such as limestone, are suspended through the action of primary combustion air distributed below the combustor floor. The gas cushion between the solids allows the particles to move freely, giving the bed a liquid-like characteristic (i.e., fluidized). FBC can occur in either atmospheric or pressurized boilers. Two fluidized bed designs can be used for atmospheric and pressurized FBC boilers: a bubbling fluidized bed or a circulating fluidized bed (CFB) (see **Exhibit 2-2**). An advantage of CFB boiler EGUs compared to PC-fired EGUs is fuel flexibility. A CFB boiler EGU can burn any rank of coal (including coal refuse), petroleum coke (a carbonaceous solid derived from oil refinery coker units or other cracking processes), and biomass without significant modifications.

The combustion temperature of a FBC boiler (1,500 to 1,650°F) is significantly lower than a PC-fired boiler (2,450 to 2,750°F), which results in lower NO_X formation and the ability to capture sulfur dioxide (SO_2) with limestone injection in the furnace. Even though the combustion temperature of a FBC boiler is low, the circulation of hot particles provides efficient heat transfer to the furnace walls and allows longer residence time for carbon combustion and limestone reaction. This results in good combustion efficiencies, comparable to PC-fired EGUs.

Atmospheric CFB boilers have successfully been scaled-up and are operating at a number of facilities throughout the world. **Exhibit 2-4** presents a simplified schematic of the major components of a CFB boiler EGU. Calcium in the sorbent combines with SO_2 gas to form calcium sulfite and sulfate solids, and solids exit the combustion chamber and flow into a hot cyclone. The cyclone separates the solids from the gases, and the solids are recycled for combustor temperature control. Heat in the flue gas exiting the hot cyclone is recovered in a series of heat recovery sections of the boiler to produce steam. The superheated steam leaving the boiler then enters the steam turbine, which powers a generator to produce electricity. Like PC-fired EGUs, CFB boilers can be used with either subcritical or supercritical steam cycles.

Currently, the capacity of CFB subcritical boilers ranges from 25 to 350 MWe. Examples of these systems include (Foster Wheeler North America Corp., 2009):

- Two 300 MWe CFB subcritical boilers at the Jacksonville Energy Authority power plant in Jacksonville, Florida. These units are capable of burning either 100% coal or 100% petroleum coke or any combination of the two.

- Three 262 MWe CFB subcritical boilers at the Turow power plant in Poland. The fuel for these boilers is lignite with moisture content of 45% by weight.

The largest atmospheric CFB boiler in operation to date is a 460 MWe unit at a power plant owned by the Polish utility company Południowy Koncern Energetyczny SA (PKE) in Lagisza, Poland (Foster Wheeler North America Corp., 2009). This unit is also the world's first supercritical CFB boiler. The primary fuel burned in the unit is Polish bituminous coal. The commercial operation of this unit demonstrates the successful integration of CFB boiler technology with supercritical boiler technology. The unit features include a vertical evaporator with supercritical steam conditions (4,000 psia, 1,050/1,075°F) and a reported overall net plant efficiency of 41.6% (HHV basis). Based on the design and operating experience with the Lagisza Power Plant, both 600 and 800 MWe size supercritical CFB boiler designs with full commercial guarantees are being offered (Foster Wheeler North America Corp., 2009).

Pressurized fluidized-bed combustion (PFBC) systems are FBC systems that operate at elevated pressures (typically pressures of 1-1.5 MPa) and produce a high-pressure gas stream at temperatures that can drive a turbine. As with atmospheric FBC, two formats are possible, one with bubbling beds, the other with a circulating configuration. Currently, all operating units use bubbling beds. In a PFBC, the combustor and hot gas cyclones are all enclosed in a pressure vessel. Both coal and sorbent (for SO_2 emissions reductions) have to be fed across the pressure boundary, and similar provision for ash removal is necessary. For hard coal (i.e., bituminous coal) applications, the coal and limestone can be crushed together, and then fed as a paste, with 25% water. As with atmospheric FBC, a combustion temperature between 1,500 to 1,650°F (800 to 900°C) has the advantage of less NO_X formation than in PC combustion. In addition, the effectiveness of a CCS system is increased due to the high pressure within the PFBC cycle and higher partial pressure of the CO_2 in the hot gas stream.

The initial or first-generation PFBC designs are based on directly burning crushed coal in the combustor. The high pressure gas is first expanded through a turbine and then heat is recovered from the turbine exhaust in a HRSG to produce steam, which is used to drive a conventional steam turbine. **Exhibit 2-5** presents a simplified schematic of the major components of a PFBC EGU. A number of demonstration projects (ranging in size from 60 to 130 MWe) were conducted during the 1990s in Japan, Spain, Sweden, the United Kingdom, the U.S., and other countries. Japanese equipment manufacturers and electric power companies have led the commercial development of PFBC technology with the construction of several commercial-scale units.

- 360 MWe PFBC unit operated by Kyushu Electric Power Company at the Karita Power Station located near Kitakyushu, Japan. The unit began commercial operation in July 2001. The unit uses a supercritical boiler and has a reported net efficiency based on test results of 41.8% HHV (Asai, 2004).

14

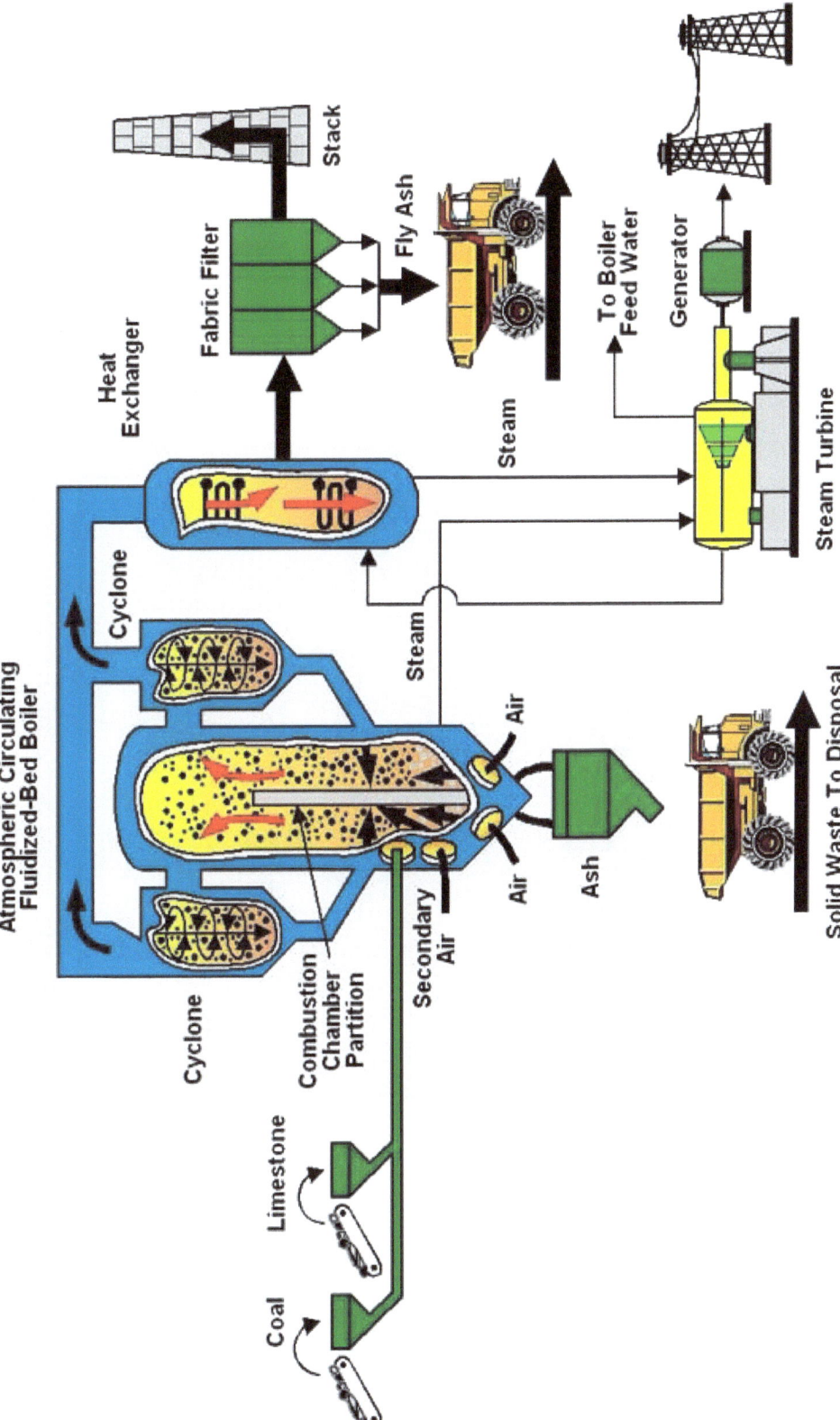

Exhibit 2-4. Simplified schematic of an atmospheric circulating fluidized-bed (CFB) boiler power plant.

Source: NETL, 2010b, *CCPI/Clean Coal Demonstrations Nucla CFB Demonstration Project, Project Fact Sheet.*

15

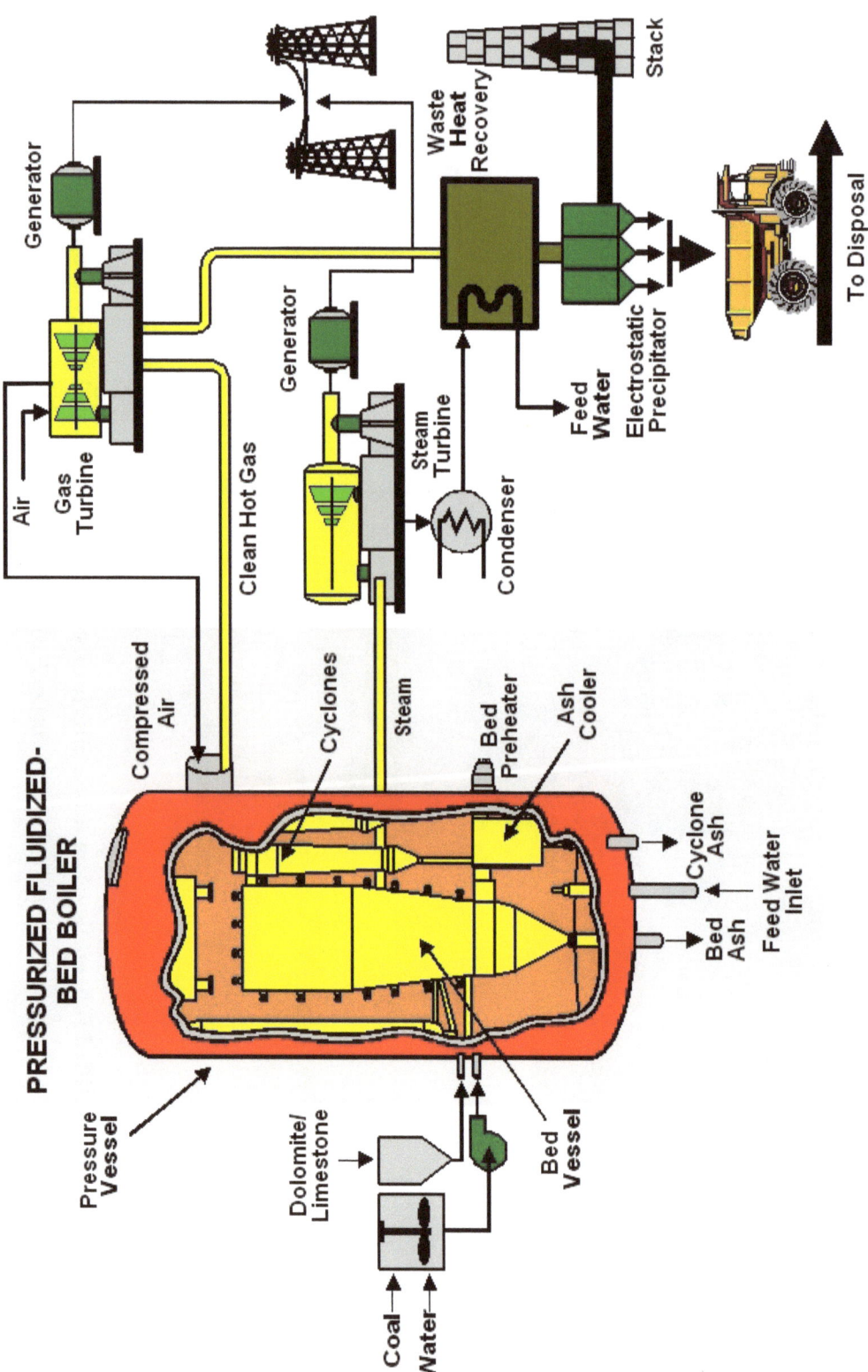

Exhibit 2-5. Simplified schematic of a pressurized fluidized-bed combustion (PFBC) power plant.

Source: NETL, 2010c, *CCPI/Clean Coal Demonstrations Tidd PFBC Demonstration Project, Project Fact Sheet.*

16

- 250 MWe PFBC unit operated by Chugoku Electric Power Co., Inc. at Osaki Power Station located near Hiroshima, Japan. Unit 1 began commercial operation in 2000. The planned construction of a second PFBC unit at the facility was cancelled in 2008

More advanced second-generation PFBC system designs use a pressurized carbonizer to first process the feed coal into fuel gas and char (solid material that remains after light gases and tar have been driven-out during the initial stage of combustion). The PFBC burns the char to produce steam and to heat combustion air for the combustion turbine. The fuel gas from the carbonizer burns in a topping combustor linked to a combustion turbine, heating the gases to the rated firing temperature of the combustion turbine. Heat is recovered from the combustion turbine exhaust in a HRSG to produce steam, which is used to drive a conventional steam turbine. These systems are also called advanced circulating pressurized fluidized-bed combustion (APFBC) combined cycle systems.

2.2.5 *Coal Gasification*

An IGCC power plant uses a coal gasification system to convert coal into a synthetic gas, which is then used as fuel in a combined cycle electric generation process. Coal is gasified by a process in which coal or a coal/water slurry is reacted at high temperature and pressure with oxygen (or air) and steam in a vessel referred to as a "gasifier" to produce a combustible gas composed of a mixture of carbon monoxide (CO) and hydrogen. This gas is often referred to as synthetic gas or syngas. Gasification processes have been developed using a variety of designs including moving bed, fluidized bed, entrained flow, and transport gasifiers. Coal gasification processes are offered by a number of companies with varying degrees of existing commercial application (NETL, 2010a). **Exhibit 2-6** presents a simplified schematic of the major components of an IGCC power plant. The hot syngas can then be processed to remove sulfur compounds, mercury, and PM before it is used to fuel a combustion turbine generator to produce electricity. The heat in the exhaust gases from the combustion turbine is recovered to generate additional steam. This steam, along with the steam produced by the gasification process, then drives a steam turbine generator to produce additional electricity.

The efficiency of an IGCC power plant is comparable to the latest advanced PC-fired and CFB EGU designs using supercritical boilers. The advantages of using IGCC technology can include greater fuel flexibility (e.g., capability to use a wider variety of coal ranks), potential improved control of PM, SO_2 emissions, and other air pollutants, with the need for fewer post-combustion control devices (e.g., almost all of the sulfur and ash in the coal can be removed once the fuel is gasified and prior to combustion), generation of less solid waste requiring disposal, and reduced water consumption when compared to an EGU using a supercritical boiler (U.S. EPA, 2006). Disadvantages of using IGCC include additional plant complexity, higher construction costs, and poorer performance at high altitude locations when compared to an EGU using a supercritical boiler. However, IGCC power plants offer the potential for lower control costs of CO_2 emissions because the CO_2 in the syngas can be removed prior to combustion. Interest by U.S. electric utilities in building new IGCC power plants is increasing because of site-specific considerations and potential cost benefits for the technology. Currently operating IGCC plants include the following U.S. and foreign plants (NETL, 2010a):

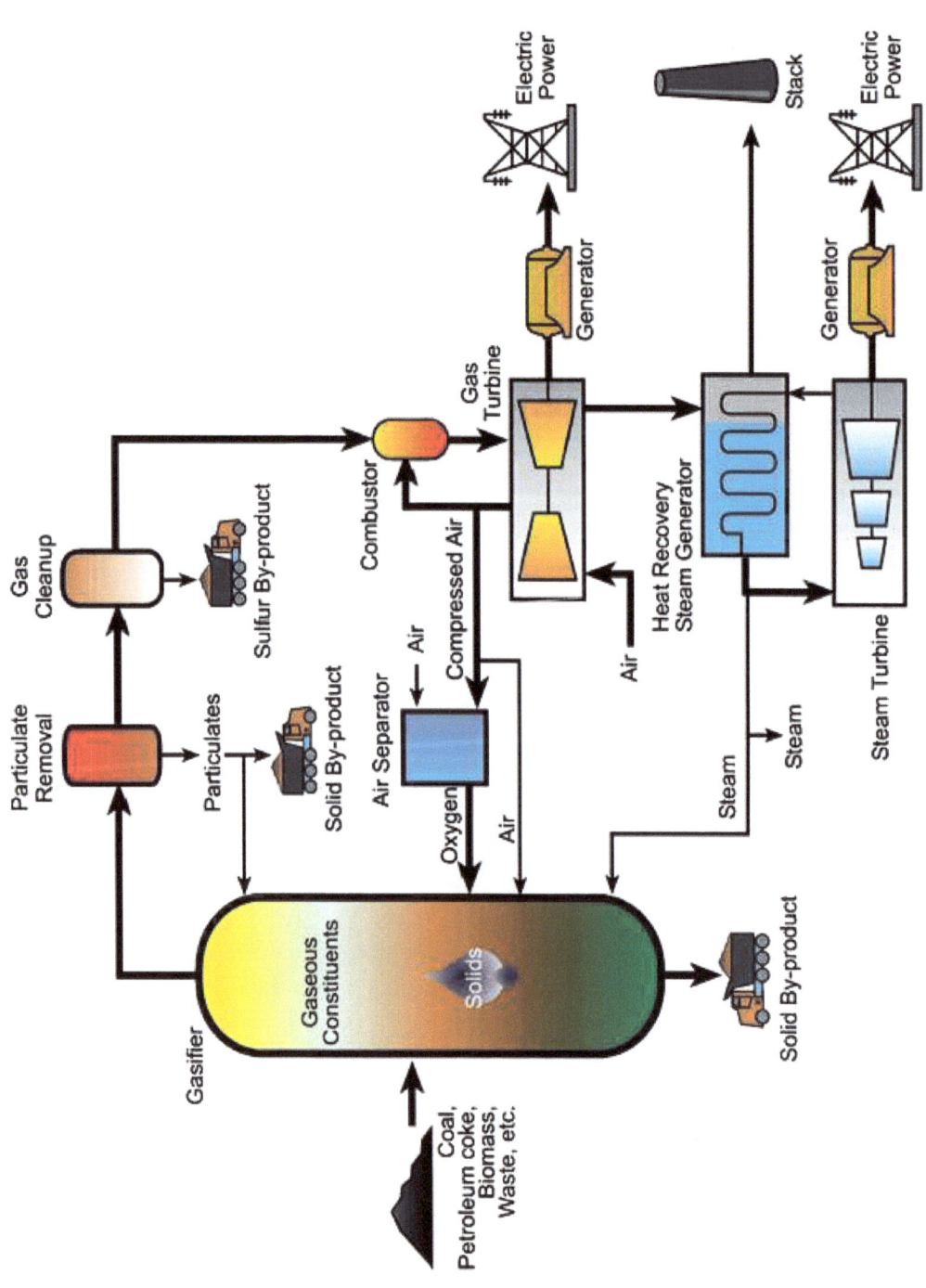

Exhibit 2-6. Simplified schematic of an integrated gasification combined cycle (IGCC) power plant.

Source: NETL, 2010a, *Overview of DOE's Gasification Program.*

- 253 MWe IGCC plant at the NUON Willem-Alexander Power Plant in Buggenum, Netherlands. The unit began operation in 1994.
- 262 MWe IGCC plant at the Duke Energy Wabash River Power Station in Indiana. The unit began operation in 1995.
- 250 MWe IGCC plant at the Tampa Electric Company (TECO) Polk Power Station in Florida. The unit began operation in 1996.
- 400 MWe IGCC plant at the SUV power plant in Vresova, Czech Republic. The unit began operation in 1996.
- 283 MWe IGCC plant at the ELCOGAS power plant in Puertollano, Spain. The unit began operation in 1998.
- 250 MWe IGCC plant at the Joban Joint Electric Power Company Nakoso Power Station in Iwaki City, Japan. The unit began operation in 2007.

Over the past 5 years, a number of larger IGCC power plant projects have been proposed by U.S. electric utility companies. Some of these IGCC projects have been indefinitely delayed or canceled because of economic and regulatory factors, such as escalating project investment costs beyond initial estimates and unresolved cost recovery issues with State public utility commissions. One commercial IGCC project currently under construction is a 630 MWe IGCC facility at the Duke Energy Edwardsport Power Station in Knox County, Indiana.

Syngas produced by coal gasification can not only be used as a fuel to generate electricity or steam but also as a basic chemical building block for a large number of petrochemical and refining products. Because of these multiple uses, future IGCC projects may include facilities that integrate electricity generation with the production of other industrial outputs such as chemical feedstocks for manufacturing operations or hydrogen fuel for vehicles.

2.3   GHG Emissions from Coal-Fired EGUs

The principal chemical constituents of coal are carbon, hydrogen, oxygen, nitrogen, sulfur, moisture, and incombustible mineral matter (i.e., ash). When coal is burned, the carbon and hydrogen are oxidized to form the primary combustion products of CO_2 and water. Other combustion products such as NO_X, SO_2, CO, and PM are formed in varying amounts.

The principal GHGs that enter the atmosphere because of human activities are CO_2, nitrous oxide (N_2O), methane (CH_4), hydrofluorocarbons (HFC's), perfluorocarbons (PCF's), and sulfur hexafluoride (SF_6). Of these, CO_2 is by far the most abundant GHG emitted from power production by coal utilization. To optimize overall efficiency for a given EGU, the unit is operated under conditions such that nearly all of the fuel carbon is converted to CO_2 during the combustion process. Methane is emitted during the mining and transport of coal but is not a significant by-product of EGU coal combustion. Fluorinated gases are not formed by coal combustion. Sulfur hexafluoride might be used at the power plant switchyard, but the switchyard is not typically considered part of the EGU.

Formation of N_2O during the combustion process results from a complex series of reactions and its formation is dependent upon many factors. However, the formation of N_2O is minimized when combustion temperatures are kept high and excess air is kept to a minimum. PC-fired EGUs are typically operated at these conditions and are not significant sources of N_2O emissions. However, FBC EGUs can have measurable N_2O emissions, resulting from the lower combustion temperatures and the use of selective noncatalytic reduction (SNCR) to reduce NO_X emissions. Operating factors impacting N_2O formation include combustion temperature, excess air, and sorbent feed rate (Korhonen, 2001). The N_2O formation resulting from SNCR depends upon the reagent used, the amount of reagent injected, and the injection temperature (Weijuan, 2007).

2.4 Factors Impacting Coal-Fired EGU CO_2 Emissions

The level of CO_2 emissions that can potentially be released from a given coal-fired EGU depends on the type of coal burned, the overall efficiency of the power generation process, and use of air pollution control devices.

2.4.1 *Impact of Coal Rank on CO_2 Emissions from EGUs*

The amount of CO_2 that potentially can be emitted from a coal-fired EGU varies depending on the coal rank burned. The amount of heat released by coal combustion depends on the amounts of carbon, hydrogen, and oxygen present in the coal and, to a lesser extent, on the sulfur content. Hence, the ratio of carbon to heat content depends on these heat-producing components of coal, and these components vary by coal rank. **Exhibit 2-7** presents a comparison of the CO_2 emissions for the average heating values of U.S. coals. The values presented in the table are arithmetic averages and assume complete combustion. Based on these averages, in general anthracite emits the largest amount of CO_2 per million Btu (MMBtu), followed by lignite, subbituminous coal, and bituminous coal. However, for a given coal rank there is variation in the CO_2 emission factor depending on the coal bed from which the coal is mined.

Exhibit 2-7. CO_2 emission factors for coal by coal rank.

Coal Rank	CO_2 Emissions per Unit of Heat Input (lbs CO_2/MMBtu)	
	U.S. Average	**Range Across States with Coal Rank Deposits**
Anthracite	227.4	227.4
Bituminous	205.3	201.3 to 211.6
Subbituminous	211.9	207.1 to 214.0
Lignite	216.3	211.7 to 220.6

Source: U.S. EIA (Hong, R. and E. Slatick, 1994).

In addition to the lower CO_2 emissions rate per unit of heat input (lbs CO_2/MMBtu), due to the inherent moisture in subbituminous and lignite coals, all else being equal a bituminous coal-fired boiler is more efficient than a corresponding boiler burning subbituminous or lignite coal. Therefore, switching from a low to a high-rank coal will tend to lower GHG emissions from the utility stack. However, overall GHG emissions might not be lowered by switching to bituminous coal. All coal mining operations release coal bed methane to the atmosphere during the mining process. Some bituminous coal reserves release significant amounts of methane, which could, in theory, offset GHG savings. Additional factors when considering overall GHG emissions include the fuel needs to mine, process, and transport the coal.

Additional solid fuels burned in EGUs include petroleum coke, biomass, and municipal solid waste (MSW). Petroleum coke has one of the highest CO_2 emissions rate (225 lb CO_2/MMBtu) of commonly used solid fuels. MSW combustors provide significant GHG reductions as an alternate to landfills (Kaplan, 2008). However, due to the difficulties associated with transporting large amounts of solid waste, MSW combustor facilities used for electrical power generation are typically limited to less than 100 MW of electrical output.

Of the gaseous and liquid fossil fuels used in steam generating units, natural gas combustion releases approximately 117 pounds of CO_2 per MMBtu, distillate oil releases 161 lb CO_2/MMBtu, and residual oil releases 174 lb CO_2/MMBtu. However, none of these fuels are typically used in new baseload steam generating units (e.g., boilers). Natural gas and distillate oil are significantly more expensive per unit heat input than coal. In addition, combustion turbines burning natural gas and distillate oil generate power more efficiently than a boiler burning natural gas and distillate oil. New baseload electric generation based on the use of either natural gas or distillate would likely use combined cycle combustion turbines. Therefore, aside from small amounts of natural gas for startup, shutdown, and potentially for combustion control, few new steam generating units are expected to burn significant quantities of either of these fuels directly in the boiler. Existing EGUs that burn natural gas and distillate oil tend to be older units that operate in a peaking or cycling mode. However, several base load coal-fired EGUs have been converted to natural gas. Natural gas-fired boilers tend to be less efficient than coal-fired boilers; however, they can startup and change loads more quickly than similar coal-fired boilers, do not typically require post combustion controls, and the fuel handling is simpler. Residual oil also tends to be more expensive than coal per unit of heat input, and because post-combustion environmental controls would still often be required, it is also not a common fuel choice for EGUs in the Lower 48 States. There has not been a new residual oil-fired EGU built in the Lower 48 States since 1981.

2.4.2 *Impact of Coal-Fired EGU Efficiency on CO_2 Emissions*

As the thermal efficiency of a coal-fired EGU is increased, less coal is burned per kilowatt-hour (kWh) generated, and there is a corresponding decrease in CO_2 and other air emissions. There is no standardized procedure for continuous on-line measurement of coal-fired EGU thermal efficiency (Peltier, 2010). However, a near approximation performed under EPA's Acid Rain Program collects heat input and gross megawatt output on an hourly basis to calculate gross heat rate. The heat input is derived from standardized continuous emission monitors, while the utility supplies gross megawatt output. The electric energy output as a fraction of the fuel energy input expressed in percentage is a commonly-used practice for reporting the efficiency of a coal-fired EGU. The greater the output of electric energy for a given amount of fuel energy input, the higher the efficiency for the electric generation process. Heat rate is another common way to express efficiency. Heat rate is expressed as the number of Btu or kJ required to generate a kWh of electricity. Lower heat rates are associated with more efficient power generating plants. Although the same basic formula is used to calculate efficiency for coal-fired EGUs, there are different methodologies for measuring the appropriate parameters. For example, the varying accuracy of the different methodologies can cause discrepancies in the measurement the heating value of the coal burned.

Efficiency can be calculated using the higher heating value (HHV) or the lower heating value (LHV) determined for the fuel. The HHV is the heating value directly determined by calorimetric measurement of the fuel in the laboratory. The LHV is calculated using a formula to account for the moisture in the fuel (i.e., subtract the energy required to vaporize the water in the coal and is thus not available to produce steam) and is a smaller value than the HHV. Consequently, the HHV efficiency for a given EGU is always lower than the corresponding LHV efficiency, because the reported heat input is larger. For bituminous coals the HHV efficiency value is typically about 2 percentage points lower than the corresponding LHV efficiency. For higher moisture subbituminous coals and lignites, the HHV efficiency is approximately 3 to 5 percentage points lower than the corresponding LHV efficiency (depending on moisture content). In engineering practice, HHV is typically used in the U.S. to express the efficiency of steam electric power plants while in Europe the practice is to use LHV.

Similarly, the electric energy output for an EGU can be expressed as either of two measured values. One value relates to the amount of total electric power generated by the EGU, or "gross output." However, a portion of this electricity must be used by the EGU facility to operate the unit, including pumps, fans, electric motors, and pollution control equipment. This in-facility electrical load, often referred to as the "parasitic load," reduces the amount of power that can be delivered to the transmission grid for distribution and sale to customers. Consequently, electric energy output is also expressed in terms of "net output," which reflects the EGU gross output minus its parasitic load.

When using efficiency to compare the effectiveness of different coal-fired EGU configurations and the applicable GHG emissions control technologies, it is important to ensure that all efficiencies are calculated using the same type of heating value (i.e., HHV or LHV) and the same type of electric energy output (i.e., gross MWh or net MWh).

Although there is a direct inverse correlation between coal-fired EGU efficiency and CO_2 emissions, other factors must be considered when comparing the effectiveness of GHG control technologies to improve the efficiency of a given coal-fired EGU. The actual overall efficiency that a given coal-fired EGU achieves is determined by the interaction of a combination of site-specific factors that impact efficiency to varying degrees. These factors include:

- *EGU thermodynamic cycle* – EGU efficiency can be significant improved by using a supercritical or ultra-supercritical steam cycle.
- *EGU coal rank and quality* – EGUs burning higher quality coals (e.g., bituminous) tend to be more efficient than EGUs burning lower quality coals (e.g., lignite).
- *EGU plant size* – The electric-generating capacity of EGUs ranges from approximately 25 to 1,300 MWe. Assuming an EGU efficiency of 33% (a typical efficiency for existing coal-fired EGUs), this corresponds to a heat input range of 250 to 13,400 MMBtu/hr. EGU efficiency generally increases with size because the boiler and steam turbine losses are lower for larger equipment. However, as equipment size increases the differences in these losses start to taper off.
- *EGU pollution control systems* – The electric power consumed by air pollution control equipment reduces the overall efficiency of the EGU.
- *EGU operating and maintenance practices* – The specific practices used by an individual electric utility company for combustion optimization, equipment maintenance, etc. can affect EGU efficiency.
- *EGU cooling system* – The temperature of the cooling water entering the condenser can have impacts on steam turbine performance. Once-through cooling systems can have an

efficiency advantage over recirculating cooling systems (e.g., cooling towers). However, once-though cooling systems typically have larger water related ecological concerns than recirculating cooling systems.

- *EGU geographic location* – The elevation and seasonal ambient temperatures at the facility site potentially may have a measureable impact on EGU efficiency. At higher elevations, air pressure is lower and less oxygen is available for combustion per unit volume of ambient air than at lower elevations. Cooler ambient temperatures theoretically could increase the overall EGU efficiency by increasing the draft pressure of the boiler flue gases and the condenser vacuum, and by increasing the efficiency of a condenser recirculating cooling system.

- *EGU load generation flexibility requirements* – Operating an EGU as a baseload unit is more efficient than operating an EGU as a load cycling unit to respond to fluctuations in customer electricity demand.

- *EGU equipment manufacturers* – The efficiency specifications of major EGU components such as boilers, turbines, and electrical generators provided by equipment manufacturers can affect EGU efficiency.

- *EGU plant components* – EGUs using the optimum number of feedwater heaters, high-efficiency electric motors, variable speed drives, better materials for heat exchangers, etc. tend to be more efficient.

Because of these factors, coal-fired EGUs that are identical in design but operated by different utility companies in different locations may have different efficiencies. Thus, the level of effectiveness of a given GHG control technology used to improve the efficiency at one coal-fired EGU facility may not necessarily directly transfer to a coal-fired EGU facility at a different location.

2.4.3 Impact of SO$_2$ Controls on Coal-Fired EGU CO$_2$ Emissions

The SO$_2$ emissions from new coal-fired EGUs, or retrofitting of an existing facility without specific SO$_2$ controls, are controlled using flue gas desulfurization (FGD) technology to remove the SO$_2$ before it is vented to the atmosphere. The selection of the type of FGD technology will impact overall GHG emissions. All FGD processes require varying amounts of electric power to operate, which contributes to the overall parasitic load of the unit. The FGD parasitic load requirements are typically between 1-2% of the gross output of the facility. In addition, some FGD processes use carbon-containing reagents (e.g., carbonates) that form CO$_2$ as a byproduct of the chemical reactions of the reagent with SO$_2$. For a typical unit, the CO$_2$ that is chemically created in a scrubber adds an additional 1% to the overall GHG emissions, but it can be as high as 3% for facilities burning high sulfur coals. However, from an overall GHG emissions standpoint the use of FGD technologies that do not form byproduct CO$_2$, such as lime-based scrubbers, do not necessarily reduce emissions. Lime is manufactured by heating limestone in the absence of oxygen to remove a molecule of CO$_2$ (CaCO$_3$ + heat → CaO + CO$_2$). Unless the CO$_2$ is sequestered at the lime production facility, overall GHG emissions will be similar. A list of FGD processes used for controlling SO$_2$ emissions from coal-fired EGUs is presented in **Exhibit 2-8,** identifying those processes that chemically form additional CO$_2$.

23

Exhibit 2-8. CO$_2$ formation from coal-fired EGU flue gas desulfurization (FGD) processes.

FGD Type	Reagent	Forms CO$_2$	Overall Reaction(s)	Reference
Wet Scrubbing	Limestone (CaCO$_3$)	yes	CaCO$_3$ + SO$_2$ + ½H$_2$O → CaSO$_3$·½H$_2$O(s) + CO$_2$; CaCO$_3$ + SO$_2$ + 2H$_2$O + ½O$_2$ → CaSO$_4$·2H$_2$O(s) + CO$_2$	Ref 1
	Magnesium-enhanced lime; Dolomitic lime [Ca(OH)$_2$•Mg(OH)$_2$]	no	10Ca(OH)$_2$ + 11SO$_2$ + Mg(OH)$_2$ → 10CaSO$_3$·½H$_2$O(s) + MgSO$_3$ + 6H$_2$O	Ref 1,4
	Dual Alkali; Sodium solution and lime	no	2NaOH + SO$_2$ → Na$_2$SO$_3$ + H$_2$O; H$_2$O + Na$_2$SO$_3$ + SO$_2$ → 2NaHSO$_3$; 2NaHSO$_3$ + Ca(OH)$_2$ → Na$_2$SO$_3$ + CaSO$_3$·½H$_2$O + 3/2H$_2$O; Na$_2$SO$_3$ + Ca(OH)$_2$ → 2NaOH + CaSO$_3$	Ref 2
	Dual Alkali (Dowa)	yes	Al$_2$O$_3$•Al$_2$(SO$_4$)$_3$ + 3SO$_2$ + 3/2O$_2$ → 2Al$_2$(SO$_4$)$_3$; 2Al$_2$(SO$_4$)$_3$ + 3CaCO$_3$ → Al$_2$O$_3$•Al$_2$(SO$_4$)$_3$ + 2CaSO$_4$(s) + 3CO$_2$	Ref 3
	Seawater	yes	2NaHCO$_3$ + SO$_2$ → Na$_2$SO$_3$ + 2CO$_2$ + H$_2$O; Na$_2$SO$_3$ + ½O$_2$ → Na$_2$SO$_4$	Ref 3
	Magnesium oxide (MgO)	no	MgO + SO$_2$ → MgSO$_3$	Ref 7
	Hydrogen Peroxide (H$_2$O$_2$)	no	H$_2$O$_2$ + SO$_2$ → H$_2$SO$_4$	Ref 3
	Sodium hydroxide (NaOH)	no	2NaOH + SO$_2$ → Na$_2$SO$_3$ + H$_2$O; Na$_2$SO$_3$ + ½O$_2$ → Na$_2$SO$_4$	Ref 3
Dry/Semi-dry Scrubbing	Hydrated calcitic lime (Ca(OH)$_2$)	no	Ca(OH)$_2$ + SO$_2$ → CaSO$_3$·½H$_2$O(s) + ½H$_2$O; Ca(OH)$_2$ + SO$_2$ + H$_2$O + ½O$_2$ → CaSO$_4$·2H$_2$O(s)	Ref 1
	Sodium bicarbonate (NaHCO$_3$)	yes	2NaHCO$_3$ + SO$_2$ → Na$_2$SO$_3$ + 2CO$_2$ + H$_2$O; 2NaHCO$_3$ + SO$_2$ + ½O$_2$ → Na$_2$SO$_4$ + 2CO$_2$ + H$_2$O	Ref 8
	Sodium sesquicarbonate (trona)	yes	2(Na$_2$CO$_3$•NaHCO$_3$•2H$_2$O) + 3SO$_2$ → 3Na$_2$SO$_3$ + 5H$_2$O + 4CO$_2$; 2(Na$_2$CO$_3$•NaHCO$_3$•2H$_2$O) + 3SO$_2$ + 3/2O$_2$ → 3Na$_2$SO$_4$ + 5H$_2$O + 4CO$_2$	Ref 8
	Sodium carbonate (soda ash, Na$_2$CO$_3$)	yes	Na$_2$CO$_3$ + SO$_2$ + ½O$_2$ → Na$_2$SO$_4$ + CO$_2$	Ref 8
	Pulverized limestone	yes	CaCO$_3$ + SO$_2$ + 2H$_2$O + ½O$_2$ → CaSO$_4$·2H$_2$O(s) + CO$_2$	Ref 1
Other Processes	Ammonia (NH$_3$)	no	SO$_2$ + 2NH$_3$ + H$_2$O + ½O$_2$ → (NH$_4$)$_2$SO$_4$	Ref 1,5
	Activated carbon	no	SO$_2$ + H$_2$O + ½O$_2$ → H$_2$SO$_4$	Ref 6

Ref 1. Srivastava, R., W. Jozewicz, and C. Singer, 2001.
Ref 2. Srivastava, R. and W. Jozewicz, 2001.
Ref 3. Davenport, 2006.
Ref 4. Benson, 2003.
Ref 5. He, 2002.
Ref 6. EPA, 2005.
Ref 7. Shand, 2009.
Ref 8. Maziuk, 2002

3. Coal-Fired EGU CO₂ Control Technologies

The development of effective and commercially viable CO_2 control technologies for coal-fired EGUs is receiving widespread attention from the utilities, technology providers, and government agencies. Some CO_2 control technologies are still in the research and development phase and are not yet ready for commercial application. Other CO_2 control technologies are being demonstrated at larger scales and are progressing towards commercial viability. This remains an active area of research and new projects and technology advances are reported routinely. The discussions of CO_2 mitigation technologies and options presented in this section are based on the development status of a given technology as described in publicly available information as of May 2010.

3.1 Coal-Fired EGU CO₂ Emissions Control Approaches

A number of technologies lowering CO_2 emissions from coal-fired EGUs are currently commercially available or under development. These control measures use one of two basic approaches to reduce the amount of CO_2 released to the atmosphere: 1) by reducing the amount of fuel used (and the amount of CO_2 formed) by improving the energy efficiency of the electrical generation process, or 2) by separating the CO_2 for long-term storage using carbon capture technology.

3.1.1 Efficiency Improvements

When the efficiency of the power generation process is increased, less coal is burned to produce the same amount of electricity. This provides the benefits of lower fuel costs and reduced air pollutant emissions (including CO_2). A number of energy efficiency technologies are available for application to both existing and new coal-fired EGU projects that can provide incremental step improvements to the overall thermal efficiency. The energy efficiency technologies with the potential to achieve the greatest improvements in electric power generation efficiency involve EGU design, equipment selection, and cost decisions that are typically incorporated during the planning and engineering design phases for a new EGU project.

3.1.2 Carbon Capture and Storage

Carbon capture and storage (CCS) involves the separation and capture of CO_2 from flue gas, or syngas in the case of IGCC. It also requires pressurization of the captured CO_2, transportation via pipeline if necessary, and injection and long-term geologic storage. Several different technologies, at varying stages of development, may be considered for the CO_2 separation and capture. Some have been demonstrated at the slip-stream or pilot-scale, while many others are still at the bench-top or laboratory stage of development.

Development of commercially viable processes for capturing CO_2 from EGUs is being funded by U.S. DOE, electric utility companies, and other organizations. These processes typically use solvents, solid sorbents, and membrane-based technologies for separating and capturing CO_2. Amine-based solvent systems are in commercial use for scrubbing CO_2 from industrial flue gases and process gases. However, solvents have yet to be applied to removing the large volumes of CO_2 that would be required for a coal-fired EGU. Solid sorbents can be used to capture CO_2 through chemical adsorption, physical adsorption, or a combination of the two effects. Membrane-based capture uses permeable or semi-permeable materials that allow for the selective transport/separation of CO_2. Oxy-combustion uses high-purity oxygen (O_2) instead of air to combust coal, producing a highly concentrated CO_2 stream that does not require a separation/capture step.

Once the CO_2 is captured, it is transported, if necessary, and stored. Geologic formations such as oil and gas reservoirs, unmineable coal seams, and underground saline formations are potential options for long-term storage. Basalt formations and organic rich shales are also being investigated for potential use as storage. Beneficial reuse (e.g., enhanced oil recovery or carbonation) is a potential alternative to strict storage that provides potential revenue to offset a portion of the CCS costs.

One recent study prepared for the U.S. DOE by the Pacific Northwest National Laboratory (PNNL, 2009) evaluated the development status of various CCS technologies. The study addressed the availability of capture processes; transportation options (CO_2 pipelines); injection technologies; and measurement, verification, and monitoring technologies. The study concluded that, in general, CCS is technically viable today. However, full-scale carbon separation and capture systems have not yet been installed and fully integrated at an EGU. The study also did not address the cost or energy requirements of implementing CCS technology. For up-to-date information on Department of Energy's National Energy Technology Laboratory's (NETL) Carbon Sequestration Program go to the NETL web site at: http://www.netl.doe.gov/technologies/carbon_seq/.

In 2010, an Interagency Task Force on Carbon Capture and Storage was established to develop a comprehensive and coordinated federal strategy to speed the commercial development and deployment of CCS technologies. The Task Force is specifically charged with proposing a plan to overcome the barriers to the widespread, cost-effective deployment of CCS within 10 years, with a goal of bringing 5 to 10 commercial demonstration projects online by 2016. As part of its work, the Task Force prepared a report that summarizes the state of CCS and identified technical and non-technical barriers to implementation. For additional information on the Task Force and its findings on CCS, go to: http://www.epa.gov/climatechange/policy/ccs_task_force.html. Because the development status of CCS technologies and their applicability to coal-fired EGUs are thoroughly discussed in the Task Force report, there will be no further discussion in this document.

3.2   Efficiency Improvements for Existing Coal-fired EGU Projects

Numerous efficiency improvements can be applied to coal-fired EGUs to increase thermal efficiency of power production (NETL, 2008, Sargent & Lundy, 2009; U.S. DOE, 2009; U.S. DOE, 2010). One specific example is the NETL study, which conducted a literature review of published articles and technical papers identifying potential efficiency improvement techniques applicable to existing coal-fired EGUs. Efficiency improvements can be expressed in different formats; they may be reported as an absolute change in overall efficiency (e.g., a change from 40% to 42% represents a 2% absolute increase). They may also be presented as the relative change in efficiency (e.g., a change from 40% to 42% is a relative change in efficiency and fuel use of 5%). The relative change in efficiency is the most consistent approach, since it corresponds to the same change in heat rate.

A summary of the findings from the NETL study is presented in **Exhibit 3-1**. The efficiency percentages were converted to a common basis so that all of the data could be compared. All of these improvements could not necessarily be implemented at each coal-fired EGU because of site-specific factors.

3.3 Efficiency Improvements for New Coal-Fired EGU Projects

3.3.1 Steam Cycle

The theoretical maximum achievable thermal efficiency achievable by an EGU using the Rankine cycle regardless of the technologies used is approximately 63% because of thermodynamic limitations and energy losses that cannot be recovered. Existing coal-fired EGUs using the Rankine cycle operate well below this limit. If the energy input to the cycle is kept constant, increasing the pressures and temperatures for the water-steam cycle will increase the output and the overall efficiency. However, a practical limitation to the higher pressure and temperatures that can be achieved in a boiler is the availability of boiler materials that can withstand these elevated conditions over an acceptable service life. The majority of existing PC-fired EGUs have subcritical boilers. Subcritical boilers typically operate at pressures of 2,400 psi (17 MPa) and at temperatures between 1,000 to 1,050°F (540 to 570°C). However, subcritical boilers can be designed to operate at steam pressures as high as 3,200 psi (22 MPa) and steam temperatures as high as 1,050°F (570°C).

The use of materials that can withstand the high-temperature and pressure of supercritical steam conditions allows for substantial improvements in efficiency for EGUs. "Supercritical" is a thermodynamic term describing the state of a substance where there is no clear distinction between the liquid and the gaseous phase (i.e., they are a homogenous fluid). Technically, the term "boiler" should not be used for a supercritical pressure steam generator, as no "boiling" actually occurs in this device, but it is common practice to use the term "Benson boiler." Supercritical EGUs typically use steam pressures of 3,500 psi (24 MPa) and steam temperatures of 1,075°F (580°C). However, supercritical boilers can be designed to operate at steam pressures as high as 3,600 psi (25 MPa) and steam temperatures as high as 1,100°F (590°C). Above this temperature and pressure the steam is sometimes called "ultra-supercritcal".

For a supercritical boiler, the feed water enters the boiler, is converted to steam, and is passed directly to the steam turbine (a supercritical boiler does not have a steam drum as shown in **Exhibit 2-3** for a subcritical boiler PC-fired EGU configuration). Because the water-steam cycle medium is a single phase fluid with homogeneous properties, there is no need to separate steam from water in a drum. Supercritical boilers operate as once-through boilers in which the water and steam generated in the furnace waterwalls passes through only once. This eliminates the need for water/steam separation in drums during operation, and allows a simpler separator to be employed during start-up conditions. Because these units do not have thick-walled steam drums, their start-up times are quicker, further enhancing efficiency and plant economics. Due to the availability of steam turbines that are designed for supercritical steam conditions, supercritical applications are presently limited to facilities of approximately 200 MWe gross output or more. Supercritical boilers are a well-established technology, and over 500 supercritical plants are currently operating worldwide (VGB, 2008).

Exhibit 3-1. Existing coal-fired EGU efficiency improvements reported for actual efficiency improvement projects.

Efficiency Improvement Technology	Description	Reported Efficiency Increase[a]
Combustion Control Optimization	Combustion controls adjust coal and air flow to optimize steam production for the steam turbine/generator set. However, combustion control for a coal-fired EGU is complex and impacts a number of important operating parameters including combustion efficiency, steam temperature, furnace slagging and fouling, and NO_X formation. The technologies include instruments that measure carbon levels in ash, coal flow rates, air flow rates, CO levels, oxygen levels, slag deposits, and burner metrics as well as advanced coal nozzles and plasma assisted coal combustion.	0.15 to 0.84%
Cooling System Heat Loss Recovery	Recover a portion of the heat loss from the warm cooling water exiting the steam condenser prior to its circulation thorough a cooling tower or discharge to a water body. The identified technologies include replacing the cooling tower fill (heat transfer surface) and tuning the cooling tower and condenser.	0.2 to 1%
Flue Gas Heat Recovery	Flue gas exit temperature from the air preheater can range from 250 to 350°F depending on the acid dew point temperature of the flue gas, which is dependent on the concentration of vapor phase sulfuric acid and moisture. For power plants equipped with wet FGD systems, the flue gas is further cooled to approximately 125°F as it is sprayed with the FGD reagent slurry. However, it may be possible to recover some of this lost energy in the flue gas to preheat boiler feedwater via use of a condensing heat exchanger.	0.3 to 1.5%
Low-rank Coal Drying	Subbituminous and lignite coals contain relatively large amounts of moisture (15 to 40%) compared to bituminous coal (less than 10%). A significant amount of the heat released during combustion of low-rank coals is used to evaporate this moisture, rather than generate steam for the turbine. As a result, boiler efficiency is typically lower for plants burning low-rank coal. The technologies include using waste heat from the flue gas and/or cooling water systems to dry low-rank coal prior to combustion.	0.1 to 1.7%
Sootblower Optimization	Sootblowers intermittently inject high velocity jets of steam or air to clean coal ash deposits from boiler tube surfaces in order to maintain adequate heat transfer. Proper control of the timing and intensity of individual sootblowers is important to maintain steam temperature and boiler efficiency. The identified technologies include intelligent or neural-network sootblowing (i.e., sootblowing in response to real-time conditions in the boiler) and detonation sootblowing.	0.1 to 0.65%
Steam Turbine Design	There are recoverable energy losses that result from the mechanical design or physical condition of the steam turbine. For example, steam turbine manufacturers have improved the design of turbine blades and steam seals which can increase both efficiency and output (i.e., steam turbine dense pack technology).	0.84 to 2.6

Source: NETL, 2008

[a] Reported efficiency improvement metrics adjusted to common basis by conversion methodology assuming individual component efficiencies for a reference plant as follows: 87% boiler efficiency, 40% turbine efficiency, 98% generator efficiency, and 6% auxiliary load. Based on these assumptions, the reference power plant has an overall efficiency of 32% and a net heat rate of 10,600 Btu/kWh. As a result, if a particular efficiency improvement method was reported to achieve a 1% point increase in boiler efficiency, it would be converted to a 0.37 % point increase in overall efficiency. Likewise, a reported 100 Btu/kWh decrease in net heat rate would be converted to a 0.30% point increase in overall efficiency.

The majority of EGUs have a single reheat cycle where the steam is first passed through the high pressure portion of the steam turbine and is then reheated in the boiler prior to passing through the remainder of the turbine. This process increases the efficiency of the EGU without increasing the maximum steam temperature. An additional steam cycle improvement that further increases efficiency is the use of a double reheat cycle, which reduces fuel use by approximately 1.5% compared to a similar EGU using a single reheat cycle (Retzlaff, 1996). The efficiency benefits of using a double reheat cycle have been recognized since the 1960s. However, the additional cost of a double reheat cycle has made a single reheat cycle typical for the majority of EGUs in the U.S.

To establish performance and cost baselines for analyzing EGU technology, NETL funded an independent assessment of the cost and performance of fossil energy power systems. The assessment specifically includes PC-fired boilers, IGCC, and natural gas-fired combined cycle systems in a consistent technical and economic manner reflecting market conditions for plants starting operation in 2010 (NETL, 2007). Performance and cost estimates were prepared for each configuration with and without CO_2 CCS. The Total Plant Cost (TPC) and Operation and Maintenance (O&M) costs for each of the cases in the study were estimated in January 2007 dollars and assumed plant construction on a generic site. The costs do not include owner cost and additional costs for special site-specific considerations at a given site.

For the PC-fired systems, the NETL study included a comparison of subcritical (2,400 psig/1,050°F/1,050°F) and supercritical (3,500 psig/1,100°F/1,100°F) PC-fired EGUs, each rated at nominal 550 MWe net capacity and firing Illinois No. 6 bituminous coal. A summary comparing the results for the subcritical unit versus the supercritical unit (without CO_2 CCS) is presented in **Exhibit 3-2**. The analysis shows an efficiency (HHV) of 36.8% for the subcritical boiler compared to 39.1% for the supercritical boiler.

Exhibit 3-2. Summary of NETL performance, cost, and CO_2 emissions comparison analysis for nominal 550 MWe PC-fired EGU burning bituminous coal by steam cycle.

Parameter	Bituminous Coal-Fired EGU	
	Subcritical Boiler	Supercritical Boiler
Gross Power Output (kWe)	583,315	580,260
Auxiliary Power Requirement (kWe)	32,780	30,110
Net Power Output (kWe)	550,445	550,150
Coal Flow Rate (lb/hr)	437,699	411,282
HHV Thermal Input (kW) Net Plant	1,496,479	1,406,161
HHV Net Efficiency (%)	36.8%	39.1%
Total Plant Cost ($ x 1,000)[a]	$852,612	$866,391
Total Plant Cost ($/kW)[a]	$1,549	1,575
Levelized Cost of Electricity (mills/kWh)[b]	64.0	63.3
CO_2 Emissions (ton/hr)	519.1	487.6
CO_2 Emissions (ton/year)	3,864,884	3,632,301
CO_2 Emissions (lb/MMBtu)	203	203
CO_2 Emissions (lb/MWh gross output)	1,780	1,681
CO_2 Emissions (lb/MWh net output)	1,886	1,773

Source: NETL, 2007. Exhibit ES-2.
[a] The NETL costs are presented as "overnight costs" in January 2007 dollars and do not include escalation, owner's costs, taxes, site specific considerations, labor incentives, etc.
[b] 10 mills are equivalent to 1 U.S. cent

Continuing research and advances in metallurgy have allowed the development of supercritical boilers capable of operating at increasingly higher temperatures and pressures, achieving increasingly higher efficiencies. Ultra-supercritical (USC) boilers designed to operate at steam conditions in excess of 4,500 psi can potentially operate at efficiencies approaching 50% (HHV). Steam temperature and pressure selection for boilers depends in part upon fuel corrosiveness, and research is focused on the development of new materials for boiler tubes and high alloy steels that minimize corrosion. There are potential concerns that temperatures above 1,100°F (590°C) while firing high-sulfur coal (such as Illinois No. 6) would result in an exponential increase of the material degradation of the highest temperature portions of the superheater and reheater due to coal ash corrosion. This could require pressure parts replacement outages every 10 to 15 years. The availability and reliability of materials required to support the elevated temperature environment for high sulfur or chlorine applications, although extensively demonstrated in the laboratory, has not been fully demonstrated commercially (NETL, 2007). Additional factors that could limit steam temperatures and pressures are the maximum values specified in the American Society of Mechanical Engineers (ASME) Boiler and Pressure Vessel Code, Section I: Power Boilers. A developer deviating from this code could have difficulty acquiring insurance, or be out of compliance with specific state code requirements.

The commercial use of ultra-supercritical technology has historically been, and continues to be, prevalent in countries outside the U.S., such as Denmark, Germany, and Japan. Ultra-supercritical boilers burning various coal ranks are being widely deployed throughout the world. Construction of the first modern ultra-supercritical EGU built in the U.S. began in 2008 at the

Southwest Electric Power Company's John W. Turk, Jr. Power Station near Texarkana, AR. This 600 MWe PC-fired facility will burn PRB subbituminous coal and is scheduled to begin operation in late 2012. Examples of ultra-supercritical PC-fired EGUs often cited as representing the currently highest efficiency operating coal-fired EGUs in the world include:

- 384 MWe ultra-supercritical PC-fired EGU for the European Vattenfall company's Nordjylland Power Station Unit 3 located near Aalborg, Denmark (Vattenfall, 2006). This power station began operation in 1998 and is a combined heat and power (CHP) facility that generates electricity and produces heat for a district heating system. The unit burns imported bituminous coal and uses seawater for cooling. Reported overall unit efficiency is 47% LHV or 45.3% HHV (International Energy Agency [IEA] Clean Coal Centre, 2007).

- 965 MWe ultra-supercritical PC-fired EGU for the German RWE Power company's Niederaussem Power Station Unit K located near Cologne, Germany (RWE Power, 2004). This unit burns lignite with a 51 to 58% moisture content and started operation in 2002. Reported unit operating efficiency is 43.2% LHV or 37% HHV (IEA Clean Coal Centre, 2007).

- Two 600 MWe ultra-supercritical PC-fired EGUs for the J-POWER (Electric Power Development Co., Ltd.) Isogo Thermal Power Station New Unit 1 and New Unit 2 located in Yokohama, Japan (J-POWER, 2009). These units burn domestic and imported bituminous coal and use seawater for cooling. New Unit 1 started operation in 2002 and has a reported unit operating efficiency of 42% LHV or 40.6% HHV (IEA Clean Coal Centre, 2007). New Unit 2 started commercial operation in July 2009. The New No. 2 Plant, has a higher efficiency due to the boosting the reheat steam temperature 18°F higher than the New Unit No. 1 to 1,148°F (J-POWER, 2009).

- 450 MWe ultra-supercritical PC-fired EGU for the Capital Power Corporation's Genesee Power Station Unit 3 located near Edmonton, Alberta, Canada (Peltier, 2005). This unit burns Alberta subbituminous coal. This unit began operation in 2005 and has a reported unit operating net efficiency or 41% LHV or 39.6% HHV (IEA Clean Coal Centre, 2007).

As part of NETL's performance and cost baseline analysis for electricity production (NETL, 2007), IGCC EGUs were analyzed for the General Electric Energy (GEE), ConocoPhillips (CoP), and Shell coal gasification processes. The IGCC cases have different gross and net power outputs than the PC-fired EGU cases because of the combustion turbine size constraint. The advanced F-class turbine used to model the IGCC cases comes in a standard size of 232 MWe when operated on syngas. Each IGCC case uses two combustion turbines for a combined gross output of 464 MWe. Additional electrical output is generated by steam turbines, with steam from the HRSGs extracting heat from the combustion turbine exhaust. Although the two combustion turbines provide 464 MWe gross output in all cases, the overall combined cycle gross output ranges from 742 to 770 MWe. The net outputs range from 623 to 640 MWe depending on the gasification process. A summary comparison of the results for the three gasification processes without CO_2 CCS is presented in **Exhibit 3-3**. Although the efficiency of the combined cycle block is approximately 50% efficient in converting the syngas to electricity, parasitic loads of the gasification process lower the net efficiency for the IGCC to 38.2 to 41.1% HHV.

Exhibit 3-3. Summary of NETL performance, cost, and CO_2 emissions comparison for an IGCC power plant by gasification process.

Parameter	Gasification Process		
	GEE Radiant	**CoP E-Gas™**	**Shell**
Gross Power Output (kWe)	770,350	742,510	748,020
Auxiliary Power Requirement (kWe)	130,100	119,140	112,170
Net Power Output (kWe)	640,250	623,370	635,850
Coal Flow Rate (lb/hr)	489,634	463,889	452,620
HHV Thermal Input (kW) Net Plant	1,674,044	1,586,023	1,547,493
HHV Efficiency (%)	38.2%	39.3%	41.1%
Total Plant Cost ($ x 1,000)[a]	1,160,919	1,078,166	1,547,483
Total Plant Cost ($/kW)[a]	1,813	1,733	1,977
Levelized Cost of Electricity (mills/kWh)[b]	78.0	75.3	80.5
CO_2 Emissions (ton/hr)	561.9	539.1	527.1
CO_2 Emissions (ton/year)	3,937,728	3,777,815	3,693,990
CO_2 Emissions (lb/MMBtu)	197	199	200
CO_2 Emissions (lb/MWh gross output)	1,459	1,452	1,409
CO_2 Emissions (lb/MWh net output)	1,755	1,730	1,658

Source: NETL, 2007. Exhibit ES-2.

[a] The NETL costs are presented as "overnight costs" in January 2007 dollars do not include escalation, owner's costs, taxes, site specific considerations, labor incentives, etc.

[b] 10 mills are equivalent to 1 U.S. cent

3.3.2   Coal Drying

Low-rank coals (lignite and subbituminous) are often utilized because the low cost per unit of heat input relative to bituminous coal and the low sulfur content. However, a major disadvantage of low-rank coals is their high moisture content, typically 25 to 40%. When this coal is burned, considerable energy is required to vaporize and heat the moisture, thus raising the heat rate of the EGU and lowering its efficiency. As fuel moisture decreases, the heating value of the fuel increases so that less coal needs to be fired to produce the same amount of electric power. Drier coal is also easier to handle, convey, and pulverize – reducing the burden on the coal-handling system. In addition, an EGU boiler designed for dried coal is smaller and has lower capital costs than a comparable EGU designed to burn coal that has not been dried. The pre-combustion drying of low-rank coals can improve the overall efficiency and several advanced coal drying technologies are or nearly are commercial available.

Great River Energy developed a coal drying technology for low-rank coals in partnership with the U.S. DOE as part of the DOE's Clean Coal Power Initiative (U.S. DOE, 2007). The technology has been successfully demonstrated on a PC-fired boiler burning lignite at the utility's Coal Creek Station in Underwood, ND. The technology is now commercially offered under the trade name DryFining™ (U.S. DOE, 2010b). The DryFining™ process passes warm cooling water from the steam turbine exhaust condenser through an air heater where ambient air is heated before being sent to a fluidized bed coal dryer. The dried coal leaving the fluidized bed is sent to a pulverizer and then to the boiler. Air leaving the fluidized bed is filtered before being vented to the atmosphere. In addition to using power plant waste heat to reduce moisture,

DryFining™ also segregates particles by density. This means a significant amount of higher density compounds containing sulfur and mercury can be sorted out and returned to the mine rather than utilized in the boiler. The end result is that more energy can be extracted from the coal while simultaneously reducing emissions of mercury, sulfur dioxide, and NO_X. At the Coal Creek Station, the process increased the energy content of the lignite from 6,200 to 7,100 Btu/lb, thereby resulting in a decrease in the fuel input into the boilers by 4% and a corresponding decrease in CO_2 emissions. Net gains in overall efficiency of 2 to 4% are reported for the process.

RWE Power in Germany is also developing a fluidized bed drying technology for lignite, called WTA (RWE Power, 2009). A fundamental difference between the two drying processes is the WTA process first mills then dries the lignite while the DryFining™ process first dries then mills the lignite. A prototype commercial-scale drying plant using the WTA process began operation in 2009 at the utility's Nederaussem Power Station site. For the WTA process, lignite is first milled to a fine particle size by hammer mills in direct series with a two-stage fluidized-bed dryer. The dried fuel exiting the stationary bed is separated from the gas stream and mixed with coarser lignite solids collected from the bottom of the dryer bed and then fed directly to the boiler. The heat needed for the drying of the fuel is supplied by external steam, which is normally taken from the turbine with the heat transfer taking place in tube bundles located inside the bed. Based on the development work completed to date of the WTA technology, the net gain in cycle efficiency is reported to be on the order of 4 percentage points, depending on the moisture content of the raw coal and the final moisture of the dried lignite.

Several other coal drying technologies are in ongoing development. One coal drying process being developed by DBAGlobal Australia Pty, Ltd., with the trade name Drycol process uses the controlled application of microwave radiation to dry coal (Graham, 2007). Coal feed stock is first separated into fine grade coal and one or more larger grades. The fine coal is loaded onto a conveyor and conveyed continuously through a microwave-energized heating chamber for drying. The fine grade coal is dried sufficiently so that when it is recombined with the larger grade coals, the moisture content of the aggregate coal is within a target moisture content range. Other coal drying technologies for low-rank coals in various stages of development include: 1) attrition milling of coal followed by air drying to produce a low-moisture coal product, 2) compressing heated, coarse crushed coal to squeeze water out , and 3) heating wet coal under pressure to approximately 480 to 570°F (APP, 2008).

To date, it has not been economic to dry subbituminous coal at the mine prior to transport to an EGU. In addition, subbituminous coal that has been dried results in increased coal dust, can spontaneously combust, and will reabsorb moisture during transport and storage. However, the development of more efficient drying technologies such as dryers using flue gas recirculation and briquetting of the dried coal to avoid spontaneous combustion and moisture reabsorption can improve the economics of upgrading low-rank coals. Descriptions of upgrading low rank coals are available from the White Energy Company (http://www.whiteenergyco.com) and Evergreen Energy (http://www.evgenergy.com/). From an overall GHG perspective, the increased EGU efficiency and decreased transportation GHG emissions would have to be compared to the energy required to dry and process the coal at the mine.

3.3.3 Boiler Feedwater Heating & Hot-Windbox

The high-pressure liquid water entering the steam generator is called feedwater. A feedwater heater is an EGU component used to pre-heat water delivered to the boiler section. Thermodynamic optimization of this cycle is important to overall EGU efficiency. In a conventional EGU, the energy used to heat the feedwater is steam extracted between the stages of the steam turbine (see **Exhibit 2-3**). Therefore, approximately a quarter of the steam that would be used to perform expansion work in the turbine (and generate power) is not utilized for that purpose. However, using other heat sources for the feedwater heater avoids the need to extract steam from the turbine allowing the steam to be used for electric power generation and increases the output of the steam cycle and potentially lowers GHG emissions. This alternate heat source can either be from an integrated solar thermal energy source or from a combustion turbine. Examples of solar thermal energy used to augment the steam cycle at combined cycle facilities include the Martin Next Generation Solar Center in Florida and the proposed Green Energy Partners/Stonewall, LLC facility in Virginia. The first coal-fired power plant to integrate solar thermal technology is the Cameo generating station in Colorado. In addition, EPRI (Electric Power Research Institute) is currently evaluating adding solar thermal energy to the Escalante and Mayo coal-fired power plants. An example of combustion turbine integration for feedwater heating is the Kettle Falls Generating Station (Schimmoller, 2003). For coal-fired boiler systems optimized to accommodate the combustion turbine exhaust, the incremental fuel efficiencies would be expected be comparable with combined cycle generation (Escosa, 2009; Stenzel). Another potential approach to integrate the use of a combustion turbine with a coal-fired steam cycle is using the turbine exhaust directly in the boiler in a hot-windbox. This involves injection of the combustion turbine exhaust directly into the boiler windbox or primary air ducts to provide an oxygen source as well as a heat source.

3.4 Combined Heat and Power Plant

Coal-fired EGUs dedicated to electric power generation and using the latest commercially available advanced technologies will generally operate at overall net efficiencies of approximately 40%. Significant amounts of energy released by coal combustion are lost during the steam condensation segment of the Rankine cycle due to heat transfer into the cooling water. In Europe, electricity is commonly generated by facilities that serve as both electricity generators and thermal energy producers for the local town or city district heating system. These combined heat and power (CHP) facilities are also known as cogeneration facilities. Operating an electric power station in a CHP mode allows recovery of some of the heat that would otherwise be rejected into cooling water, improving the overall efficiency of energy utilization. In applying CHP to an existing or new EGU, the temperature of the cooling water is normally not high enough to meet the requirements for most district heating or industrial process applications. In these cases, steam would be extracted at an elevated pressure and temperature from an intermediate stage of the steam turbine and then used for district or process heating. This results in a decrease in the total electric power generation from the EGU. However, the overall fuel efficiency of CHP is higher than if electricity and steam were generated separately.

Because electricity can be transmitted over long distances, electric power plants can be located in remote areas as well as urban areas. However, thermal energy cannot be effectively transported over extended distances. This limits the practicality of incorporating a CHP mode into many electric power plant designs. The EGU needs to be located in close proximity to either a district energy system or an industrial facility with a significant and steady thermal demand. There are a number of examples; however, where industrial facilities have collocated with existing or new coal power plants in order to have access to reliable, low cost steam:

- DuPont's titanium dioxide plant in Johnsonville, Tennessee, is located next to TVA's Johnsonville power plant and buys high pressure process steam from the 1,200 MW facility. The power plant is comprised of 10 coal-fired boiler steam turbine units; the DuPont plant uses steam extracted from Units 1 through 4. Providing steam to the DuPont facility at the required process pressures reduces overall output of the power plant by 50 MW.

- Blue Flint Ethanol in Underwood, North Dakota, is a 50 million gallon per year dry mill ethanol producer located next to Great River Energy's 1,160 MW Coal Creek lignite-fired power plant. Starting operations in 2007, the ethanol facility purchases approximately 100,000 pounds per hour of medium pressure steam extracted from the power plant.

- Goodland Energy Center in Goodland, Kansas, is a small 22 MW coal-fired power plant that is supplying steam to a 20 million gallon per year ethanol plant and a 12 million gallon per year biodiesel plant, both co-located with the power plant. The power plant and the ethanol plant both started construction in 2006.

3.5 Oxygen Combustion

Oxygen combustion (oxy-combustion, oxy-firing or oxy-fuel) is an emerging technology applicable to either new or existing EGUs. The advantage offered by this technology is its potential for CO_2 emissions control because it produces a concentrated (nearly pure) CO_2 exhaust gas stream that requires minimal post-combustion clean-up prior to compression, transportation, and injection for long term storage. The basic concept of oxy-combustion is to use a mixture of oxygen (or oxygen-enriched air) and recycled flue gas (containing mostly CO_2) in place of ambient air for coal combustion. The resulting flue gas contains primarily CO_2 and water vapor with smaller amounts of oxygen, nitrogen, SO_2, and NO_X. Consequently, the flue gas can be processed relatively easily to further purify the CO_2 (if necessary) for use in enhanced oil or gas recovery or for geological storage.

An oxy-combustion power plant consists of an air separation unit (ASU), an EGU with O_2-blown combustion, and a CO_2 treatment unit. The conventional ASU is a cryogenic process that has a significant energy requirement. However, alternative oxygen separation methods are being researched for possible commercial scale development. These alternative methods include ion transport membranes (ITM), ceramic autothermal recovery, oxygen transport membranes, and chemical looping (UARG, 2008). Oxygen is mixed with recirculated flue gas to create a mixture of O_2 and CO_2 (and some H_2O) which is used as the source of combustion oxidant instead of ambient air. The absence of air nitrogen produces a flue gas stream with a high concentration of CO_2.

Several research institutes are focusing on laboratory- and pilot-scale testing of oxy-fuel combustion (Levasseur, 2009). Pilot test programs currently are being conducted for European Enhanced Capture of CO_2 (ENCAP) program and the Advanced Development of the Coal-Fired Oxyfuel Process with CO_2 Separation (ADECOS) program. Additional research and development programs are being conducted, including:

- A 30 MW oxy-firing pilot plant at the Schwarze Pumpe station in Spremberg, Germany. This plant is the first complete oxy-combustion unit that includes the integrated system from the air separation unit to the gas purification and compression systems. The CO_2 will be compressed and liquefied for storage experiments to be conducted.

- A 32 MW oxy-firing demonstration project in France retrofitting an existing boiler to natural gas oxy-combustion. The captured CO_2 will be transported through an approximately 19 mile long pipeline and stored in a depleted gas field in Lacq, South of France.

- A comprehensive test program using the 15 MW tangentially-fired Boiler Simulation Facility and 15 MW Industrial Scale Test Facility operated by Alstom Power, Inc., in Windsor, CT. Testing is being conducted to assess a broad range of oxy-combustion design options. Project partners include the U.S. DOE, the Illinois Clean Coal Institute, and 10 electric utility companies.

4. Coal-Fired EGU Technology Alternatives Analysis

There is no one best available coal-fired EGU technology universally applicable to all EGU projects. The coal-fired EGU technology alternatives most suitable for a given project must be evaluated on a site-specific basis. An evaluation for a new facility would include the use of carbon capture and storage and the most efficient technologies (e.g., ultra-supercritical steam conditions, IGCC, pressurized fluidized bed), double steam reheat, coal drying, FGD technology, and CHP.

4.1 Site-Specific Coal-Fired EGU Technology Alternatives Analysis Example

This section summarizes the results for analyses prepared in support of an air permit application for a new 830 MW supercritical PC-fired EGU to be built by the Consumers Energy Company at their Karn-Weadock Generating Station in Bay County, Michigan. The EGU is designed to burn PRB subbituminous coal, but can also mix bituminous coal based on supply or price variations. A supercritical PC boiler with steam turbine throttle pressure of 3,805 psia and superheat and single reheat temperatures of 1,100 °F was proposed by the Consumers Energy Company.

The air permit application was initially submitted to the State of Michigan Department of Natural Resources and Environment (DNRE) in 2007. As an amendment to the permit application for the Consumers Energy project, the DNRE requested that Consumers Energy Company prepare and submit a top-down BACT analysis for IGCC plant as an alternative to building the supercritical PC-fired EGU. The supercritical PC-fired EGU air emissions controls included selective catalytic reduction (SCR), baghouse, wet flue gas desulfurization, hydrated lime injection, and activated carbon injection. The IGCC plant air emissions controls included SCR, Selexol or Sulfinol, black water handling equipment, and sulfur-impregnated activated carbon.

To estimate the IGCC plant capital costs, the permit applicant developed a premium factor for constructing an IGCC plant compared to a similar capacity supercritical PC-fired EGU based on information collected by the permit applicant including reported costs of other IGCC projects. This capital cost premium factor was then applied to the estimated costs for the supercritical PC-fired EGU. Additional cost estimates were made for fuel, water consumption, waste disposal, operation and maintenance, and pollutant allowance purchases required for each facility type. The reference IGCC plant capital costs were estimated to be approximately 24% higher than the supercritical PC-fired EGU. The analysis estimated that the cost of electricity generation from an IGCC unit would be approximately 37% higher for the IGCC unit than for the supercritical PC-fired EGU. The projected cost of generation for the supercritical PC-fired EGU was $60/MWh compared to $95/MWh for the IGCC plant. The cost estimates prepared for the analysis are summarized in **Exhibit 4-1**.

Exhibit 4-1. Supercritical PC-fired EGU and IGCC plant cost comparison Summary prepared for Consumers Energy EGU project.

Cost Parameter[a]	800 MWe net Supercritical PC-fired EGU	800 MWe net IGCC Plant	Difference (IGCC vs. PC)
Capital Costs ($)	$2,671,916,111	$3,526,039,934	$854,123,823
Annualized Costs ($/yr)	$273,871,401	$361,419,093	$87,547,692
Fuel Cost ($/yr)	$117,624,738	$130,679,946	$13,055,208
Cooling Water Consumption Cost ($/yr)	$5,166,656	$3,114,268	(-$2,052,388)
Waste Disposal Cost ($/yr)	$316,937	$201,042	(-$115,895)
Operating and Maintenance Costs ($/yr)	$59,307,369	$84,336,386	$25,029,017
Total Annual Cost ($/yr)	**$456,287,101**	**$579,750,735**	**$123,463,634**
Annualized Cost of SO_2 Allowances ($/yr)	$2,442,975	$610,531	(-$1,832,445)
Annualized Cost of NO_x Allowances ($/yr)	$4,132,417	$2,611,787	(-$1,520,630)

Source: Consumers Energy Company, 2008.

[a] Costs are presented in September 2007 dollars and include owner's and financing costs.

At the request of the State air permitting authority for the project, the permit applicant prepared and submitted a second electric generation alternatives analysis for the Consumers Energy project in June 2009 (Consumers Energy Company, 2009). As part of this alternatives analysis, a comparison of the various coal-fired EGU technologies was presented with respect to air emissions from coal combustion including CO_2. The ultra-supercritical PC-fired EGU has the lowest projected heat rate of the analyzed technologies. Of the technologies without carbon capture and storage, it also has the lowest GHG emissions rate. In December 2009, the Michigan Department of Environmental Quality issued the construction permit based on the proposed supercritical PC boiler. The results are summarized in **Exhibit 4-2**.

Exhibit 4-2. Coal-fired EGU technology alternatives cost comparison summary prepared for Consumers Energy EGU project.

Coal-Fired EGU Technology	40-yr. BusBar Cost Excluding CO_2 Cost[a]	40-yr. BusBar Cost Including CO_2 Cost[a]	Availability	Efficiency-Heat Rate (Btu/kWh)	CO_2 Emission Rate (tons/MWh)
Supercritical PC-fired 830 MW	$97 per MWh	$133 per MWh	86% to 92%	9,134	0.94
Subcritical PC-fired	$101 per MWh	$136 per MWh	84% to 89%	9,407	0.97
Subcritical CFB boiler	$108 per MWh	$145 per MWh	87%	9,798	1.01
Supercritical CFB boiler	$108 per MWh	$144 per MWh	87%	9,508	0.98
IGCC plant	$128 per MWh	$162 per MWh	70% to 81%	9,490	0.93
Ultra-supercritical PC-fired	$98 per MWh	$133 per MWh	91%	9,019	0.93
Supercritical PC-fired with Carbon Capture and Storage (CCS)	$135 per MWh	$139 per MWh	86% to 91%	10,836	0.11
Supercritical PC-fired 500 MW	$104 per MWh	$140 per MWh	86% to 92%	9,134	

Source: Consumers Energy Company, 2009.

[a] Busbar costs is the cost to generate the power leaving the plant (beyond the generator but prior to the voltage transformation point in the plant switchyard) and include all plant fixed costs (including all costs associated with the capital investment), fuel costs, operating and maintenance costs, emissions costs, interconnection costs, and transmission system upgrade costs. A busbar cost excluding CO_2 costs assumes a CO_2 tax or cap-and-trade program has not been implemented. A busbar cost including CO_2 costs assumes a CO_2 tax cost of $22/ton beginning in 2012 and rising to $53/ton by 2025.

4.2 EPA GHG Mitigation Database

The EPA Office of Research & Development (ORD) is collecting information regarding CO_2 mitigation measures applicable to coal-fired EGUs for compilation in a publicly-accessible GHG Mitigation Database. Version 1 of this database is expected to be released to the public in late summer 2010. The database is a tool that provides information of both commercially available technologies, as well as emerging technologies that are being demonstrated at larger scales for commercial viability.

EPA Contact

Christian Fellner
U.S. EPA
OAQPS/SPPD/ESG
Mail Code D243-02
Research Triangle Park, NC 27711
Phone: 919-541-4003
fellner.christian@epa.gov

Nick Hutson, Ph.D.
U.S. EPA
OAQPS/SPPD/ESG (detail)
Mail Code E305-01
Research Triangle Park, NC 27711
Phone: 919-541-2968
hutson.nick@epa.gov

References

Internet Web Site addresses cited for individual references were available as of the date of publication of this document.

Asai, Akihisu, et al. 2004. *System Outline and Operational Status of Karita Power Station New Unit 1 (PFBC)*. JSME International Journal, Series B. Vol. 47, No. 2, 2004. Pp. 193-199. Available at: <http://www.jstage.jst.go.jp/article/jsmeb/47/2/193/_pdf>.

Asia-Pacific Partnership on Clean Development and Climate (APP). 2008. *Brown Coal Drying Technologies*. APP Joint Meeting of Cleaner Fossil Energy and Power Generation and Transmission Task Forces. April 1, 2008. Available at: <http://www.asiapacificpartnership.org/pdf/CFE/meeting_melbourne/SurveyofBrownCoalDryingTechnologies-Godfrey.pdf>.

Benson, Lewis, et al. 2003. *Control of Sulfur Dioxide and Sulfur Trioxide Using By-Product of a Magnesium-Enhanced Lime FGD System*. Presented at ICAC Forum '03 Multi-Pollutant Emission Controls & Strategies, Multi-Pollutant Emission Controls & Strategies, Nashville, TN . October 14-15, 2003. Available at: <http://www.carmeusena.com/files/files/techpapersreports/carmeuse_icac_20forum_2003.pdf>.

Benson, Lewis. *New Magnesium-Enhanced Lime Flue Gas Desulfurization Process*. Technical Paper Report, Carmeuse North America. Pittsburgh, PA. Available at: <http://www.carmeusena.com/files/files/TechPapersReports/fgd_new_magnesium.pdf>.

Consumers Energy Company. 2008. *Amendment to Application No. 341-07 Consumers Energy Company – Bay Count Top-Down BACT Analysis for IGCC*. April 2008. Consumers Energy Company, Jackson, MI. Available at: <http://www.deq.state.mi.us/aps/downloads/permits/CFPP/2007/341-07/Top-Down%20BACT%20Analysis%20for%20IGCC.pdf >.

Consumers Energy Company. 2009. *Balanced Energy Initiative: Electric Generation Alternatives Analysis*. June 2009. Consumers Energy Company, Jackson, MI. Available at: <http://www.deq.state.mi.us/aps/downloads/permits/PubNotice/341-07/AlternativesAnalysis.pdf >.

Davenport, W.G., et al. 2006. *Sulfuric Acid Manufacture*. Southern African Pyrometallurgy. South African Institute of Mining and Metallurgy, Johannesburg, 5-8 March 5-8, 2006.

Escosa, Jesús M. and Luis M. Romeo. 2009. *Optimizing CO_2 Avoided Cost by Means of Repowering*. Applied Energy, 86 (2009) 2351–2358.

Foster Wheeler North America Corp. 2009. *Utility CFB Goes Supercritical - Foster Wheeler's Lagisza 460 MWe Operating Experience and New 600 - 800 MWe Designs*. Prepared by James Utt, Arto Hotta, and Stephen Goidich, Foster Wheeler North America Corp., Clinton, NJ, for presentation at Coal-Gen 2009, Charlotte, NC, August 19-21, 2009. Available at: <http://www.fwc.com/publications/tech_papers/files/TP_CFB_09_12.pdf>.

Graham, James. 2007. *Microwaves for Coal Quality Improvement: The Drycol Project*. DBAGlobal Australia, Milton Queensland, Australia. Presented at the SACPS/International Pittsburgh Coal Conference 2007, Johannesburg, South Africa, September 10-14, 2007.

Available at: <http://www.drycol.com/downloads/Drycol%20Paper%20ACPS-1%20060608.pdf>.

He, Boshu, et al. 2002. *Temperature Impact on SO₂ Removal Efficiency by Ammonia Gas Scrubbing*, Energy Conversion and Management 44:2175-2188. http://www.sciencedirect.com/science?_ob=ArticleURL&_udi=B6V02-3WRC6DR-3&_user=775537&_origUdi=B6V2P-47CY482-G&_fmt=high&_coverDate=07%2F31%2F1998&_rdoc=1&_orig=article&_acct=C0000429 38&_version=1&_urlVersion=0&_userid=775537&md5=e2354e5058cba8f383722059ffecd1 cd

Hong, B.D. and E.R. Slatick. 1994. *Carbon Dioxide Emission Factors for Coal*. Originally published in U.S. Energy Information Administration, Quarterly Coal Report, January-April 1994, DOE/EIA-0121(94/Q1). Washington, DC, August 1994, pp. 1-8. Available at: <http://www.eia.doe.gov/cneaf/coal/quarterly/co2_article/co2.html>.

IEA Clean Coal Centre. 2007. *G8 Case Studies by the IEA Clean Coal Centre*. Prepared by Colin Henderson, IEA Clean Coal Centre, London, United Kingdom for presentation at Third International Conference on Clean Coal Technologies for Our Future, Sotacarbo Coal Research Centre, Carbonia Sardinia, Italy. May 15 -17, 2007. Available at:<http://www.iea-coal.org/publishor/system/component_view.asp?LogDocId=81704&PhyDocId=6360>.

J-POWER (Electric Power Development Co., Ltd.). 2009. *Annual Report 2009*. Electric Power Development Co., Ltd., Tokyo, Japan. October 2009. Available at: <http://www.jpower.co.jp/english/ir/pdf/2009.pdf>.

Kaplan, P. Ozge, et al. 2008. *Is It Better To Burn or Bury Waste for Clean Electricity Generation?*. Environmental Science & Technology, Vol. 43, No. 6, 2009, pp. 1711-1717.

Korhonen, S., et al. 2001. *Methane and Nitrous Oxide Emissions in the Finnish Energy Production*. Fortum Tech-4615.

Levasseur, A. A., et al.. 2009. *Alstom's Oxy-Firing Technology Development and Demonstration - Near Term CO₂ Solutions*. Presented at the 34th International Technical Conference on Clean Coal & Fuel Systems, Clearwater, FL. May 31 - June 4, 2009. Available at: <http://www.netl.doe.gov/technologies/coalpower/ewr/co2/pubs/5290%20Alstom%20oxy-combustion%20paper%20Clearwater%20jun09.pdf>.

Maziuk, John and John Kumm. 2002. *Comparison of Dry Injection Acid-Gas Control Technologies*. Presented at the 95th Annual Conference and Exhibition of the Air and Waste Management Association, Baltimore, MD. June 23-27, 2002.

Michigan Department of Natural Resources and Environment. 2009. PTI Application No. 341-07. December 29, 2009. Available at: <http://www.deq.state.mi.us/aps/downloads/permits/CFPP/2007/341-07/341-07.htm>.

National Energy Technology Laboratory (NETL), 2007. *Cost and Performance Baseline for Fossil Energy Plants, Volume 1: Bituminous Coal and Natural Gas to Electricity, Revision 1*. DOE/NETL-2007/1281. U.S. Department of Energy, National Energy Technology Laboratory, Pittsburgh, PA. August 2007. Available at: <http://www.netl.doe.gov/energy-analyses/pubs/Bituminous%20Baseline_Final%20Report.pdf>,

National Energy Technology Laboratory (NETL), 2008. *Reducing CO$_2$ Emissions by Improving the Efficiency of the Existing Coal-fired Power Plant Fleet*, DOE/NETL-2008/1329. U.S. Department of Energy, National Energy Technology Laboratory, Pittsburgh, PA. July 23, 2008. Available at: <http://www.netl.doe.gov/energy-analyses/pubs/CFPP%20Efficiency-FINAL.pdf>.

National Energy Technology Laboratory (NETL), 2010a. *Overview of DOE's Gasification Program.* Presentation by Jenny Tennant, Technology Manager, Gasification, U.S. U.S. Department of Energy, National Energy Technology Laboratory, Pittsburgh, PA. January 25, 2010. Available at: <http://www.netl.doe.gov/technologies/coalpower/gasification/pubs/pdf/DOE%20Gasification%20Program%20Overview%202010%2001-25%20v1v.pdf>.

National Energy Technology Laboratory (NETL), 2010b. *CCPI/Clean Coal Demonstrations Nucla CFB Demonstration Project, Project Fact Sheet.* U.S. Department of Energy, National Energy Technology Laboratory, Pittsburgh, PA. Accessed June 21, 2010. Available at: <http://www.netl.doe.gov/technologies/coalpower/cctc/summaries/nucla/nuclademo.html>.

National Energy Technology Laboratory (NETL), 2010c. *CCPI/Clean Coal Demonstrations Tidd PFBC Demonstration Project, Project Fact Sheet.* U.S. Department of Energy, National Energy Technology Laboratory, Pittsburgh, PA. Accessed June 21, 2010. Available at: <http://www.google.com/imgres?imgurl=http://www.netl.doe.gov/technologies/coalpower/cctc/summaries/tidd/images/tidd_plant_bw.jpg&imgrefurl=http://www.netl.doe.gov/technologies/coalpower/cctc/summaries/tidd/tidddemo.html&usg=__PMP6wAdGSJtmGws-CiNKKCwryUc=&h=>

Pacific Northwest National Laboratory (PNNL), 2009. *An Assessment of the Commercial Availability of Carbon Dioxide Capture and Storage Technologies as of June 2009*, PNNL-18520. Pacific Northwest National Laboratory, Richland, WA. June 2009. http://www.pnl.gov/main/publications/external/technical_reports/PNNL-18520.pdf>.

Peltier, Robert. 2005. *Genesee Phase 3, Edmonton, Alberta, Canada.* Power. July/August 2005. Available at: <http://www.epcor.ca/SiteCollectionDocuments/Corporate/pdfs/publications%20and%20newsletters/plattspower05.pdf>.

Peltier, Robert. 2010. *Plant Efficiency: Begin with the Right Definitions.* Power. February 1, 2010. Available at: <http://www.powermag.com/gas/Plant-Efficiency-Begin-with-the-Right-Definitions_2432.html>.

Retzlaff, Klaus M and W. Anthony Ruegger, 1996. *Steam Turbines for Ultrasupercritical Power Plants.* GE Power Generation, GER-3945A. Available at: <http://www.gepower.com/prod_serv/products/tech_docs/en/downloads/ger3945a.pdf>.

RWE Power. 2004. *Niederaussem Power Plant: A Plant Full Of Energy.* RWE Power Aktiengesellschaf, Essen, Germany. October 2004. Available at: <http://www.debriv.de/tools/download.php?filedata=1218532199.pdf&filename=Kraftwerk%20Niederaussem%20(englisch).pdf&mimetype=application/pdf>.

RWE Power. 2009. *WTA Technology: A Modern Process for Treating and Drying Lignite.* RWE Power Aktiengesellschaf, Essen, Germany. February 2009. Available at:

<http://www.rwe.com/web/cms/mediablob/en/88166/data/183490/36906/rwe/innovations/power-generation/coal-innovation-centre/fluidized-bed-drying/download-wta-en.pdf>.

Sargent & Lundy. 2009. *Coal-Fired Power Plant Heat Rate Reductions*. Report Number SL-009597. Sargent & Lundy, Chicago, Il. January 22, 2009. Available at: <http://www.epa.gov/airmarkt/resource/docs/coalfired.pdf>.

Schimmoller, Brian, 2003. *Avista Kettle Falls*. Power Engineering. October, 2003. Available at: <http://www.powergenworldwide.com/index/display/articledisplay/189185/articles/power-engineering/volume-107/issue-10/departments/managing-the-plant/avista-kettle-falls.html>

Shand, Mark A., 2009. *A New Look at SO$_2$ Removal*. April 28, 2009. Available at: <http://www.risiinfo.com/technologyarchives/chemicalsnews>.

Srivastava, Ravi K, Wojciech Jozewicz and Carl Singer, 2001. *SO$_2$ Scrubbing Technologies: A Review*. Environmental Progress, 20(4):219-28.

Srivastava, Ravi K and Wojciech Jozewicz. 2001. *Flue Gas Desulfurization: The State of the Art*, Journal of the Air & Waste Management Association, 51:1676-88.

Stenzel, William C., et al. *Repowering Existing Fossil Steam Plants*. December 11, 1997. Available at: <http://soapp.epri.com/press/default.htm>.

Utility Air Regulatory Group (UARG). 2008. *A Review of Carbon Capture and Sequestration (CCS) Technology*, Prepared by J. Edward Cichanowicz, Saratoga, CA, December 2008.

U.S. Department of Energy (DOE), 1996. *Electric Utility Engineer's FGD Manual, Vol. 1--FGD Process Design*. U.S. Department of Energy, Office of Fossil Energy, Morgantown, WV. March 1996. Available at: < http://www.netl.doe.gov/technologies/coalpower/ewr/pubs/fgdmanual_vol1.pdf>.

U.S. Department of Energy (DOE). 2007. *Clean Coal Technology Power Plant Optimization Demonstration Projects Topical Report Number 25*. September 2007. Available at: <http://www.netl.doe.gov/technologies/coalpower/cctc/topicalreports/pdfs/topical25.pdf>

U.S. Department of Energy (DOE), 2009. *Technical Workshop Report: Opportunities to Improve the Efficiency of Existing Coal-Fired Power Plants*. Workshop sponsored by the U.S. Department of Energy and National Energy Technology Laboratory. Rosemont, IL. July 15-16, 2009. Available at: <http://www.netl.doe.gov/energy-analyses/pubs/NETL%20Power%20Plant%20Efficiency%20Workshop%20Report%20Final.pdf>.

U.S. Department of Energy (DOE), 2010. *Technical Workshop Report: Improving the Efficiency of Coal-fired Power Plants in the United States*. Workshop sponsored by the U.S. Department of Energy and National Energy Technology Laboratory. Baltimore, MD. February 24-25, 2010. Available at: <http://www.netl.doe.gov/energy-analyses/refshelf/detail.asp?pubID=306>.

U.S. Department of Energy (DOE). 2010b. *Innovative Drying Technology Extracts More Energy from High Moisture Coal*. Fossil Energy Techline. March 11, 2010. Available at: <http://fossil.energy.gov/news/techlines/2010/10006-CCPI_Technology_Goes_Commercial.html>.

U.S. Energy Information Administration (U.S. EIA). 2008. *EIA-923 (Schedule 2) - Monthly Utility and Nonutility Fuel Receipts and Fuel Quality Data (2008 Final Data).* U.S. Energy Information Administration, Office of Coal, Nuclear, Electric and Alternate Fuels, U.S. Department of Energy, Washington, DC 20585. Accessed June 21, 2010. Available at: <http://www.eia.doe.gov/cneaf/electricity/page/eia423.html>.

U.S. Energy Information Administration (U.S. EIA). 2010. *Electric Power Annual 2008.* DOE/EIA-0348(2008). U.S. Energy Information Administration, Office of Coal, Nuclear, Electric and Alternate Fuels, U.S. Department of Energy, Washington, DC 20585. January, 2010. Available at: <http://www.eia.doe.gov/cneaf/electricity/epa/epa_sum.html >.

U.S. Environmental Protection Agency (EPA). 2001. *Database of information collected in the Electric Utility Steam Generating Unit Mercury Emissions Information Collection Effort.* OMB Control No. 2060-0396. U.S. Environmental Protection Agency, Office of Air Quality Planning and Standards, Research Triangle Park, NC. April 2001.

U.S. Environmental Protection Agency (EPA). 2005. *Multipollutant Emission Control Technology Options for Coal-fired Power Plants.* EPA. EPA-600/R-05/034. U.S. Environmental Protection Agency, National Risk Management Research Laboratory. Research Triangle Park, NC. March 2005. Available at: <http://www.epa.gov/airmarkt/resource/docs/multipreport2005.pdf>.

U.S. Environmental Protection Agency (EPA). 2006. *Environmental Footprints and Costs of Coat-Based Integrated Gasification Combined Cycle and Pulverized Coal Technologies.* EPA. EPA-600/R-06/006. U.S. Environmental Protection Agency, National Risk Management Research Laboratory. Research Triangle Park, NC. July 2006. Available at: <http://www.epa.gov/oar/caaac/coaltech/2007_01_epaigcc.pdf>.

Vattenfall. 2006. *Nordjylland Power Station: The World's Most Efficient Coal-Fired CHP Plant.* Vattenfall, A/S Copenhagen, Denmark. Available at: <http://www.vattenfall.dk/da/file/nordjyllanduk080121_7841588.pdf>.

VGB Powertech and Evonik Energy Services (2008). Supercritical and ultra supercritical technology. In *Power Plant Performance Reporting and Improvement under the Provision of the Indian Energy Conservation Act: Output 1.1: Best practice performance monitoring, analysis of performance procedures, software and analytical tools, measuring instrumentation, guidelines or best practice manuals and newest trends.* (Annexure XIV). Indo German Energy Programme. Retrieved from <http://www.emt-india.net/PowerPlantComponent/Output1.1/Output1.1.pdf>.

Weijuan, Yang, et al. 2007. *Nitrous Oxide Formation and Emission in Selective Non-Catalytic Reduction.* Energy Power Engineering China, 228-232.